THE SYMPHONY

THE SYMPHONY

VOLUME ONE

Haydn to Dvořák

*

EDITED BY
ROBERT SIMPSON

DRAKE PUBLISHERS INC NEW YORK

ISBN 87749-244-1

PUBLISHED IN 1972 BY
DRAKE PUBLISHERS INC
381 PARK AVENUE SOUTH
NEW YORK, N.Y. 10016

PRINTED IN GREAT BRITAIN

IN MEMORIAM

JULIUS HARRISON

CONTENTS

INTRODUCTION

Robert Simpson

*

ONE often hears musicians praising some work as 'genuinely symphonic'; if, however, they are asked what they mean by this, most are hard put to it to give a very precise reply. The subject needs definition. We must know what we are going to include in a book of this kind, and why. It would be simple enough to take all composers on trust, without looking further than the titles of their works, to include a work just because it is called a symphony and bears a distinguished signature, and so to avoid trouble. But the fact that musicians do speak of 'genuinely' symphonic music, in terms however nebulous, and criticize some symphonies as being such in name only, is solid evidence that there is some essence recognized by those who discuss it at all. There are, of course, many present-day critics and composers who do not discuss it, who say that the whole idea is dead. This makes it all the more necessary to examine the situation, to find out what 'symphonic' really means, so that we can see whether it is an exhausted possibility, or whether it is something that can still fire the imagination. And even if we were to reach the conclusion that such a way of writing music is no longer possible, we must still try to discover what it is, for it has generated some of the greatest and most durable music in existence.

As Harold Truscott has admirably shown in the first chapter of this book, the origins of the symphony as we know it were in a new attitude to tonality, or key. In the third decade of the eighteenth century, tonality began to be felt in a new way, a change of key being an event that thrust the original key away over the horizon, so that it, for the time being at any rate, disappeared. Tonality is a difficult thing to describe in words, impossible in fact. Yet few people fail to understand it instinctively or, partly, by unconsciously acquired habit. Only a tone-deaf person could remain undisturbed if a simple hymn tune were to end in the wrong key. The right tonic chord is expected, demanded by us all. We also feel the rightness or wrongness of the harmonies in the middle of the tune, as the different notes of the melody are supported by various related chords. Now imagine such a process greatly extended, so that instead of a simple tune we have a much larger composition whose intermediate harmonies are stretched out into periods, each containing all sorts of activity. We can hear that, like the hymn tune, the piece possesses a sense of tonal direction, and when we reach the end, we feel the same completeness and finality. Any characteristic Bach prelude will illustrate this. The extension is such that what was originally felt as a close succession of harmonies is now so much spread out, and so enlivened by figuration or contrapuntal invention, that it seems like a drift through a series of keys. But the whole has still as static a feeling about it as the hymn tune. When we get to the end we have the idea that we have not really moved at all, but simply stood looking in various directions.

Each key has its retinue of related chords that can be used placidly without disturbing the original one. Imagine now a Bach prelude in a particular key, say A major. Get a musician to transpose it for you to E flat, and you are in a different world altogether (nothing to do with so-called 'key-colour'). The internal relations are the same, but the sound of E flat is dramatic and strange after the sound of A. These two keys are as remote from each other as possible, and there is no need to be a trained musician to receive the full impact of the sensation. E flat virtually blots out the original sensation of A, once the first wrench is over. Now it would be possible to thrust the original key over the horizon without a wrench, by means of a modulation – a raising of tension by means of various harmonies (some of them dissonances) that eventually relaxes into the new key. A modulation from A to E flat would be an extreme one, and one can more easily modulate to much nearer regions. We have seen that a Bach prelude more often than not sticks to related harmonies; its first point of rest is very likely to be on *dominant* harmony, that is to say the chord a fifth higher than the original tonic. The dominant demands a return to the tonic:

I

In this case, one tonic chord of C major is enough to satisfy the simple demand left by the unfinished quotation. The G major chord is merely a harmony in the course of a melody. If Bach arrives on it half-way through the course of an elaborate piece, the rest of the piece will find its way back, normally through an exquisite series of steps.

Suppose we treat this dominant not as a chord but as a key and we *modulate* to it, creating enough tension in the intervening process to thrust the original key out of earshot. This is what happened in the early eighteenth century, and it gave rise to a whole new world of music, essentially dramatic in feeling and dynamic in movement. The first great idea that arose out of it was the sonata principle, on which most new chamber and orchestral music was founded. If the original key is genuinely supplanted by a new one, the process of returning home again has to be pretty muscular; it gave rise to the so-called development section of a sonata movement, with its kaleidoscopic, restless changes of key. The return is itself a dramatic incident (even if it is quietly and smoothly effected), and the whole point of the recapitulation is that it has to insist upon the original tonality with much more weight than Bach would find necessary. Sometimes even the weight of the recapitulation is insufficient; then there has to be a *coda*, to reinforce it.

Countless symphonies, quartets and sonatas begin with a movement of this type. The early quartets of Haydn do not differ in any fundamental respect from his early symphonies, and the gradual divergence of his chamber and orchestral music is no divergence of musical streams – it is a gradual realization of the essential difference between the capabilities of the small intimate group and those of the increasingly 'public' orchestra. Haydn and Mozart put their best into both; in the field of chamber music the string quartet and quintet proved to be the means of conveying the highest degree of concentration. The name 'symphony' became attached to orchestral works aiming at the same kind of density and significance. And not only density; variety within a required unity was always tacitly regarded as vital, variety both of movement and character. No really great symphony lacks this internal variety, achieved within the scope of a concentrated sense of unity. This is also true of great quartets; but the quartet is so-called by a purely utilitarian, statistical name. If it is a bad, loose, or rhapsodical composition, it is still undeniably a quartet. But a symphony cannot be called such if it wants certain qualities. The name itself, whatever its primitive origins, has come to mean a work for orchestra in which the composer has obeyed and mastered (a paradox that really does mean something, for once) not a set of rules but a body of principles, or standards.

The difference between the kind of musical thought in, say, Tchaikovsky's *Francesca da Rimini* and the first movement of his Fourth symphony is not merely one of degree; it is a basic difference in kind. The one is episodic, essentially static; the other is organic, and essentially dynamic. It is no use saying that this difference resides in the fact that the movement of the symphony is in sonata form. Rules are not the root of the matter; it is easy enough to find episodic, unsymphonic movements that stick haplessly to the sonata plan (there are some by Tchaikovsky). Few musicians nowadays would be so rigid as to insist that a symphony shall deploy sonata form – there are plenty of real symphonies that have little or nothing to do with it. But (and it is tremendously important to realize this) they live by the principles of which sonata form was the first great manifestation. The symphony and the sonata arose from a new revelation about tonality; a 'breakthrough' (to use a currently fashionable word) that is as space flight is to gravity-bound air travel, and almost strictly analogous.

We cannot now put off any longer an attempt to list (however haltingly) those elements of music a composer must master if he is to write a true symphony. They must be put in the most general terms possible, for this is not, being an artistic phenomenon, a matter for the straight-jacket.

(1) The fusion of diverse elements into an organic whole. The composer must be able to create a wide range of movement and character, shape and colour, even mood and atmosphere, within severe and powerfully concentrated limits. A great symphony embraces all kinds of musical movement, from one extreme to the other. From this arises:

(2) The continuous control of pace. Even within a single movement the control of pace must be absolute; an abrupt change of motion that might be acceptable for itself in a ballet will require subsequent justification in a symphony. In a great master it will have the thrilling effect of imperiously commanding justification, as it does in the first E major outburst in the first movement of Nielsen's Fourth symphony (see Vol. 2, Ex. 41). A movement in a single pervasive tempo must show a continuous mastery of *composed* flexible pace; the listener must have the sense that it is the composer and not the conductor who is controlling this. In the first movement of the 'Eroica', though it is an unbroken *Allegro con brio*, there is far more variety of motion than in, say, Liszt's *Les Préludes*, which requires the conductor to make crashing changes of gear from time to time.

(3) The reserves of strength necessary to achieve (1) and (2) are such as to express size. A true symphony is always big in its power of suggestion, even if its physical dimensions are small. Consider any of Haydn's middle-period symphonies, Beethoven's Eighth, or Sibelius's Seventh.* A symphony may possess great dimensions but these alone are not enough; Beethoven's Seventh is gigantic in its implications, which

* I regard this work as more genuinely symphonic than does Harold Truscott (see Vol. 2, p. 97).

dwarf many a monster of sheer bulk. The mere fact of this special kind of mastery is enough to invest with a distinctive sense of power and stature even light and gay symphonies such as Mendelssohn's 'Italian', and there is more of such quality in Prokofiev's *Classical Symphony* than in his much more aggressive and expansive *Scythian Suite*.

(4) In the first place it was the dynamic treatment of tonality that made all this possible; it was a reaction against the tonal passivity of earlier music. The important thing to remember about it is that the revolt was vital and constructive, positive – it threw away nothing, not even counterpoint. In the introduction to Volume 2, I have to describe another kind of revolt, a negative one.

(5) Perhaps the basic observation one can make about true symphony is that it is active in all possible ways. That is to say that in creating it the composer must never allow any prime element of the music (rhythm, melody, harmony, tonality) to seem to die, so that artificial respiration becomes necessary. In a piece of ballet or purely picturesque music temporary failure of one of these elements need not matter seriously – the effect may even be compelling or beguiling. But in a symphony such a failure amounts to a lapse; even when movement is temporarily denied to such an ingredient as, say, tonality (as is often imperative) there must always be an awareness that the movement can be restored at will, without disrupting the continuity. In such a case the composer, if for example he leaves the bass holding a long pedal note, must *know* what he is going to do with it, and that what he will do will be positive. No evasions are tolerable in the attempt to achieve the highest state of organization of which music is capable. In the orchestral field we have long applied the term 'symphony' to such an endeavour.

Questions of what should or should not be included in a book of this nature arise more in respect of the second than the first volume, and will be discussed in the other introduction. In the meantime I would like to make it clear that so far as individual contributors are concerned, I have left each free to tackle his subject in his own way, in the belief that the upshot would be livelier and more varied than the effect of a rigid format. The demands of different subjects cause different approaches. Hans Keller, for example, has commented upon every single one of Mozart's symphonies, for this is a subject not fully dealt with elsewhere. Basil Lam in his penetrating observations about Beethoven has largely dispensed with music quotations, knowing that the works are familiar and pocket scores and records easily obtainable, and he has concerned himself with those profound generalities that routine analysis precludes. Some of the chapters, indeed, are not light reading, though they are I think very readable. We have not assumed that our readers are children – rather that they wish to reach to the heart of the matter and are prepared for a burst of rewarding effort here and there. Only one chapter has been reprinted (though slightly revised) from the previous Pelican *Symphony* – Humphrey Searle's treatment of Liszt. To all the contributors I offer my thanks, and to Julius Harrison, who died before he could see his chapters through the press, this publication is dedicated in affectionate memory.

I

JOSEPH HAYDN

(1732–1809)

AND THE RISE OF THE CLASSICAL SYMPHONY

Harold Truscott

*

SYMPHONY means literally 'ensemble', but we are concerned with one particular application of the word, and that the most important and the most familiar. Many accounts of the gradual emergence of the classical symphony (including classical sonata style in general) from earlier music are available, tracing it from the motet in several polyphonic and homophonic sections to the sonata da chiesa, through the French Overture of Lully and the Italian Overture of Alessandro Scarlatti, who began to call this operatic introduction *Sinfonia*, to the classical symphony itself. There are resemblances between the last two items. The Scarlatti overture or *Sinfonia avanti l'opera* especially, with its order of movements as quick, slow, quick, has a likeness to the early symphony; but it is a superficial one. And, in fact, most of these resemblances are very superficial. It is probably true that some of the surface characteristics of the symphony grew from some of the earlier music. But when we compare an early overture with a genuine symphony the differences far outweigh the similarities. The most we can say is that the earlier overture, especially the Scarlatti type, gradually lost some of its characteristics and took on new ones, but this is not enough to account for *the* essential difference that we find.

First, most earlier large scale music was predominantly contrapuntal; the early classical* music is not. And this

* Or pre-classical, as it is sometimes called, somewhat ambiguously, since 'Baroque' and early classical music overlap in the early eighteenth century. 'Rococo' is another term often used, and is equally misleading.

change of style came suddenly, almost overnight. It brought with it great problems, on the keyboard and in music for the groups of instruments used then variously for chamber and orchestral music. All the early classical composers, Italian as well as German, had been brought up in the prevailing contrapuntal style; therefore there must have been some great compulsion to impel them to change this style at the cost of immense trouble and with a result far cruder, to begin with, than anything in the prevailing polyphonic style.

Now it is true that if one examines seventeenth century music (including, for convenience, that of Bach and Handel), one can find a lot of surface resemblances between that music and early classical music. But they are similarities which are easier to see than to hear. If one looks at the Prelude to Bach's F major 'English' suite:

2

it will appear to have a surface resemblance to a sonata movement by Haydn – the one in F, for instance, beginning:

3

The Bach movement is large and it certainly has more themes than the Haydn; one of them has a surface likeness to a 'second subject', and there is a similarity of texture, too. It 'goes through different keys'; so does Haydn. Where, then, is the difference, if there is one? The answer is to hear the two movements. There will scarcely be any question then of connecting them. What can be heard is this: the Bach runs predominantly in consistent counterpoint, and moves from harmony to harmony, touching its tonic F at intervals, so that everything is related to that tonic. F as the centre of the whole is never in question. Even the 'second subject' is arrived at by a phrase moving from the tonic, F, to its dominant, and again from there to *its* dominant,

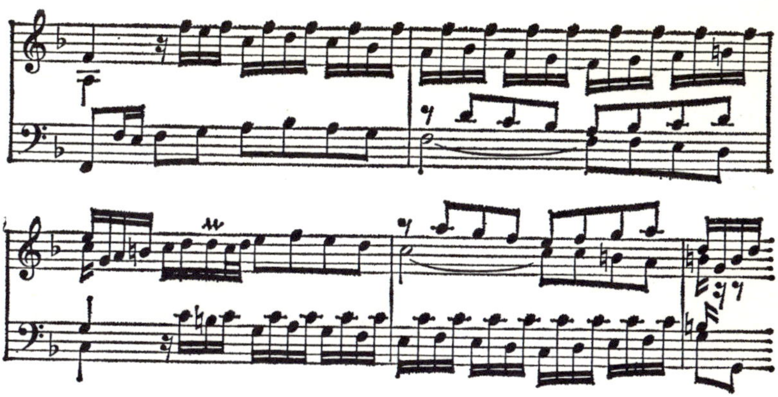
4

so that all is still in direct line with the scale of F; no foreign element has been introduced, except the chromatic B natural, which is not enough to change a key although it may be enough to change a harmony. Harmony after harmony follows in this gradual connected way and when the harmonic tour is finished and the music rests firmly on its tonic again there is absolutely no sense of having returned to something which has been absent; it has been present as the governing force all the time. Another comparison is with a walk round the outside of a cathedral, so that the same features are continually seen at different angles.

To call Bach's harmonic moves 'modulations' is to confuse them with the sort of thing which actually happens in the Haydn movement. 'The key of the dominant' is a worn phrase, most misleading because it refers to an impossibility. Either your piece moves on to the dominant (which exists as an aural fact only in relation to an existing tonic) or it has completely left the initial key and moved to an entirely different one whose tonic in notation is the same as the original dominant. Notation is all the connexion there is between them. Haydn's exposition does not end in 'the key of the dominant'; it ends in the key of C major, which is a totally different thing, in *sound*. The original tonic F is forgotten. This is what is meant by a real change of key, and it must be experienced to understand the upheaval caused by its arrival in music at the beginning of the eighteenth century. It explains why J. S. Bach did not feel attracted by the course which his sons were taking: it would have meant abandoning everything that was in his own nature.

Now the different effect of the counterpoint in both the Bach and Haydn movements can be appreciated. Although the Haydn movement begins with smooth running overlapping counterpoint, and this texture appears to go on throughout the movement, there is, in fact, a subtle change in the style only a few bars after the start of the piece. Haydn's transition theme,

keeps the semiquavers going, but they cease to be contrapuntal and become simply broken chords; in other words, his transition passage is purely harmonic, not contrapuntal at all, although it preserves the rhythmic movement of the opening counterpoint very cleverly. The change occurs, it will be noticed, as the music approaches the specifically tonal part of the exposition, the change of key. Haydn's counterpoint is very simple wherever it occurs in this piece, but it is real, and he has carefully separated the two styles, dividing them between passages which are settled harmonically or tonally and passages where key change is taking place. His movement is as different in sound from the Bach as it could be.

I do not know when this discovery was made, nor who made it; nor, I am sure, does anyone else. It does not matter; it was made. My guess would be that it was made by one or more, probably several people independently, doodling on keyboard instruments; and once made, it could not be ignored. It caused trouble from the beginning, simply because it took time to discover exactly what sort of music was fitted to make the best use of this device. Melody was used to begin with, because that is what its earliest practitioners were used to; counterpoint was used also, for the same reason. And it would not do. Therefore counterpoint was dropped and melody replaced with themes – and the very crudest of themes, at that; mostly, little more than an arpeggio up the tonic chord and down the dominant – to assert what had never needed this specific assertion before.

The distinction between the simple harmonic move from tonic to dominant or tonic to mediant, within one key, and a complete change of key is a real one; the two things are utterly different. The first is lyrical; that is to say, it is entirely on one harmonic plane, governed throughout by one tonic, and causes no dramatic complexities. The second is a sundering, an antagonistic division in sound between an area governed by one tonic and a similar area governed by another tonic. And such a process generates conflict; conflict is the basic element of drama; and what classical music eventually resolved itself into is drama; not a succession of dramatic effects but drama, continuous and working itself out to a climax. However much in later sonata music the drama may rely to some extent on melodic elements, this tonal conflict is the root of it all. No wonder that it took the earliest classical composers a long time to sort out exactly the right treatment for this astonishing device.

From this central fact of real key change came a great many consequences. First, counterpoint was dropped, simply because it slowed down the action and was an equal conflicting interest with appreciation of the tonal structure. As a result the continuo became dispensable. It was used for a while out of habit – Monn uses it in his earlier symphonies, although it is not at all necessary. Its only reason for existing had been to hold together the threads of harmony while solo instruments disported themselves contrapuntally. Hitherto there had been very little distinction between orchestral and chamber music – both used the continuo and the 'cello doubling the bass line; around this nucleus, including strings in the case of the orchestra, was built up whatever group of instruments was wanted, or was available. But there was no really stable orchestra, and the main difference between the two types of music was simply in larger and smaller groups of instruments. But now the continuo had lost its original reason for existing, counterpoint having been replaced by block harmony, the group of instruments began to supply its own continuo, at first on oboes and horns or horns alone, later adding trumpets, and so came into being the nucleus of the modern, stabilized orchestra – another far-reaching result of this vital tonal discovery.

This, then, is the cradle of classical music: a clash of real tonalities, new then in music and still fresh today if really heard. From this came almost every important development in European music for most of the next two centuries. As far as we can tell, this music began in Italy, with such composers as Rinaldo di Capua, born in 1715, and G. B. Sanmartini (or Sammartini), who was born in 1700 or 1701* and numbered Gluck among his pupils. Sammartini was the first to use a device later seized upon by many classical composers – that of making a quick move to the home dominant, insisting on it for a few bars, and starting a definite second subject there. This is the elementary mistake of taking a note still prominently associated with the existing key and assuming it to be the tonic of a new key; but Sammartini was not satisfied with this process. Usually, in his maturer works, he designs his second subject so that it works through very chromatic harmonies until it has lost all touch with the original key and so comes later to a definitely established second key. He is genuinely the father of the difficult art of real key change; and how difficult it is may be measured by the large number of outstanding composers who have failed to master it.

Sammartini has been credited with the first attempts to write a clearly defined second subject. This wrongly puts the emphasis on thematic contrast. What he did was to make a clearly defined second key, colliding with and antagonistic to the first; sometimes this occurs after his 'second subject' has begun. He seems also to have been the first to realize that if the two subjects were to be recapitulated both in the main key of the movement, something really imaginative could be made of the passage between them; that, in fact, this involved not merely the 'slight changes' so beloved of writers of programme notes, but something the reverse of the original transition passage, permitting an exploration of the main key not possible in the exposition. Some of the music written by Sammartini as a result of this realization is not far below that of Haydn and Mozart at their best.

The influence of this Italian music reached Vienna, stimulating the first important school of symphonic composition.

* His birth has been put as early as 1693.

From here eventually many composers dispersed throughout Europe, to Paris, Hamburg, London, and to Mannheim. Many of them were of Bohemian nationality. One of the finest of the Viennese group was Georg Matthias Monn, born in 1717. It is revealing to examine one of his contrapuntal works, a Sonata (or Partita) in A major, against one of his symphonies, and find the completely different treatment accorded each. The Partita begins with a normal solemn introduction called 'Overture', with the characteristic dotted rhythm:

6

It comes to a halt on the dominant of A and leads to a fugue in A in which the treatment is precisely that of any fugue by a seventeenth century composer – a series of harmonic moves round the tonic. This is particularly interesting as the work is scored for first and second violins with bass and continuo, so that a good deal of the actual fugal texture depends on the continuo:

7

The fugue is succeeded by an *Andante* in A minor, which is a violin melody supported by the other instruments; it rarely leaves the tonic or dominant chords, but includes one pungent piece of chromaticism in its seventeen bars:

8

After this we have a minuet and trio, major and minor respectively, and a finale in A major, again lyrical in tonality (that is, in one key throughout). The style of the whole is a good specimen of the late North German *French Overture*, and from this work alone one would never suspect its composer of having anything to do with the classical symphony.

However, it is as a classical symphonist that he is known to us, and in such works his methods are very different. At times he relies far more on melody than most early classical symphonists, as in the first movement of a wonderful little symphony in B major, but his methods usually more resemble those of the later Mannheim school, although he was never actively concerned with the German group. A symphony in E flat, one of his mature works, shows us his ways quite clearly. It begins with a plain theme of the tonic and dominant variety in 5 (not 4) bar phrases:

9

The strings only are shown here; his orchestra included two oboes and two horns, with a continuo. The style is obviously purely homophonic, with the repeated-note bass and the Alberti-type* viola part. Monn spends, in proportion to the length of his movement, a large number of bars effecting his key change, the bass first simply rising twice up the scale from E flat to the dominant, then sinking again from E flat, but with an A natural this time:

* To Domenico Alberti (1710–40) is given, rightly or wrongly, the credit for devising a type of keyboard bass line formed of repeated broken chords, intended to suggest the *sostenuto* that keyboard instruments, other than the organ, could not give in the normal way.

When his second subject comes, if one can separate from his continuous thematic line one theme for this position, he seeks to emphasize B flat by using A flat as a chromatic note, an extremely effective diatonic use of chromaticism:

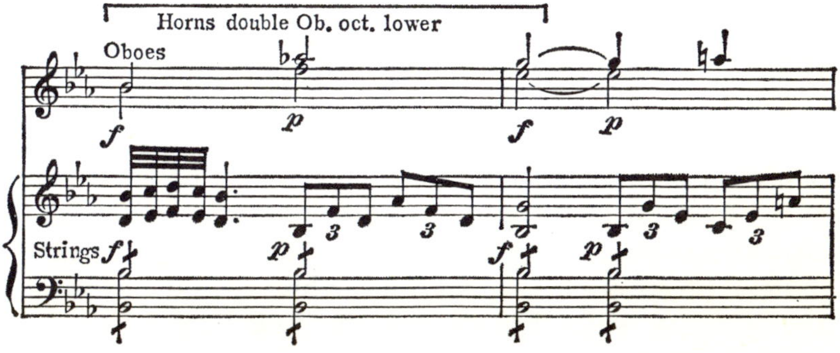

11

Even now he is not satisfied and only at the end of his exposition does he succeed in finally ramming his key home to his satisfaction:

12

Here we have an echo of Sammartini's procedure, but on a larger scale. Monn's development is short, only twelve bars, using the beginning of Ex. 9 and expanding this with some of his transitional material, but his presentation of this main theme in the recapitulation is so different as to remove any sense of monotony. He gives it in one of the longest and most difficult pieces of horn writing in the history of the classical symphony, of which I quote only the beginning:

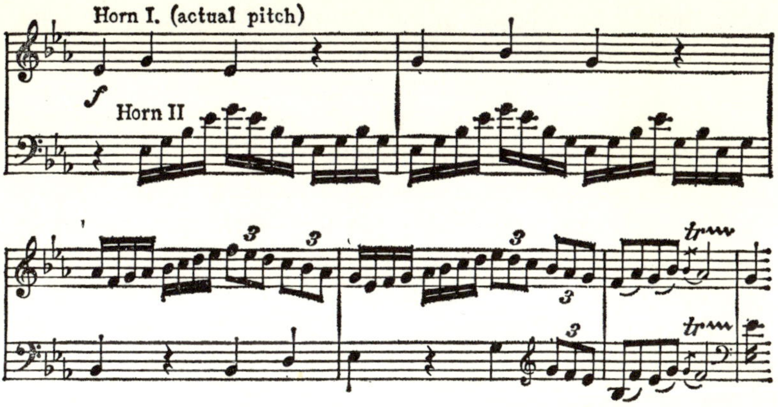

13

His management of the reversed transition necessary to keep the music now in E flat throws a wonderful retrospective light on the tonal events of his exposition. And here is a movement in which it is obvious that almost the whole of its composer's concern is tonal, as it was not in the Partita.

Monn's contribution to the development of the symphony is fine and influential. He has worthy colleagues, in Wagenseil, born in 1715, and later the Bohemian Vanhal, born in 1739, whose double recapitulations on totally different key structures marked a further step forward.

Naturally, symphonic composition developed the orchestra. A good orchestra grew up in Vienna but the most famous was that in Mannheim, which drew musicians from all over Europe. It was brought to its finest state by a Bohemian, Johann Wenzel Stamitz, born in 1717, a magnificent all-round musician, the first real conductor, one of the best of all the early symphonists and father of two others. The strings were recognized now as the mainstay of orchestral tone and, to begin with, the only wind instruments were two oboes and two horns. At Mannheim, however, flutes, clarinets as required, and bassoons were added, and later trumpets and drums as well. The Mannheim orchestra and group of composers flourished under a music-loving Elector, Karl Theodor, and they were as fortunate in him as Haydn was later in Prince Esterhazy. By the time Stamitz went to

Mannheim in 1745 most of the symphony's teething troubles were well under control. Stamitz stabilized it, perhaps a little too much, and so we begin to get the typical Mannheim symphonic opening, with the full orchestra asserting the tonic *forte*, a quiet answer usually involving the oboe or flute, solid repeated-note basses, and a feature which obviously grew from the orchestra to the music rather than in reverse: the famous Mannheim crescendo; in addition, under Jommelli's* indirect influence, there began to appear a much greater variety of dynamics, to which the eighteenth century English musical historian Burney pays tribute.

Stamitz usually gave the major musical interest to the strings in his own symphonies, and made use of the wind in beautiful supporting parts. His dovetailing of first and second violins was an intelligent anticipation of much later orchestral string writing. His son Carl, as gifted as his father, after relying at first on his father's methods, gradually relaxed them. He achieved a finer continuity of phrasing, and in places seems even to anticipate Mozart at his best. It is possible that Mozart was specially influenced by Carl as well as by Christian Bach. But the most notable product of Mannheim was Franz Beck, born there about 1730 and a pupil of Johann Stamitz. In many of his twenty-four symphonies there is music of striking concentration and his harmonic invention is often prophetic. All these composers are now overshadowed by Haydn and Mozart, yet they have something to offer which cannot be found in the work of their greater successors.

Another consequence of the achievement of real key change was a new and muscular sense of speed. To the contrapuntal composers *Allegro* meant music in a lively mood but very little quicker than a normal *Allegretto;* it could not move fast because contrapuntal detail had to come through. The composers in the new style found themselves compelled to think more consciously in terms of phrase-lengths *as such*, which, naturally, were at first somewhat

* Niccolò Jommelli (1714–1774) was Kapellmeister to the Duke of Württemberg for sixteen years from 1753.

regular. What creeps into classical sonata and symphonic music is a tendency to augment the prevailing dramatic atmosphere by inserting irregular phrases of three or six bars, sometimes phrases of five bars or a four and a half, as in the extraordinary composer, Josef Starzer, perhaps the most adventurous of all the Viennese group. Such departures from the normal began first of all to appear in passages of transition from one key to another. Key change is, in a sense, destructive but it is so in order to be, on a much larger scale, constructive. And so we arrive at a new and rapid pace, which would not have been conceivable in earlier music. The connexion between the quick, slow, quick of the Scarlatti overture and the early classical symphonies is only apparent, and depends upon assuming the Baroque use of *Allegro* to mean the same thing as the classical usage. Not until Schubert did the pace begin to slow down again, but never again to anything like the *Allegro* of the contrapuntists, or for the same reason.

When Haydn arrived, he found a congenial air to breathe. His early compositions, mostly short church settings, show him half-way between a style based on the counterpoint he had learnt and one that dallied with some vague symphonic methods. The *Missa Brevis* in F, written in 1749 or 1750, when he was eighteen, to which he added wind parts in later life, is an excellent example of this delightful but somewhat restricted style of writing. He is usually thought to have been fired to write in the new style by perusing and conceiving great admiration for some of Carl Philipp Emanuel Bach's sonatas and symphonies; the admiration was genuine, although there is some reason to believe that Emanuel Bach complained later than Haydn did not accord his work the consideration he (Bach) felt was its due. However, it is much more likely that Haydn imbibed this fresh spring of new music from the work of the Viennese classical composers, since it was in Vienna that he grew up from the age of eight.* He certainly knew Wagenseil. If he did absorb any

* There is some support for this belief in part of a letter by Haydn quoted by Karl Geiringer on p. 32 of his biography of the composer:

influence from C. P. E. Bach he transformed it. Emanuel Bach's musical personality is attractive. It shows itself through extended ornamentation and a compromise between homophonic and latent contrapuntal writing in his keyboard sonatas, and a tendency to throw out exciting dramatic gestures in his symphonies. These gestures remain gestures, similar to those an actor or an orator might indulge in for emphasis. Of Haydn's objective concentration on essentials Bach's music shows no trace. The one way in which Emanuel Bach may have influenced Haydn strongly is through his magnificent scoring. Few of the Mannheim group's most prominent characteristics show themselves in Haydn's work, but there is some Monn to be found, and a great deal of Wagenseil and Josef Starzer.

The chronology of Haydn's compositions between 1750 and 1760 is uncertain, although Mr Robbins Landon's monumental book on the symphonies has gone a long way towards establishing it. There was a set of *Divertimenti*, composed variously at Vienna and Lukavec during this decade, some pieces for wind band (1759–60) and, most important, a set of six *Scherzandi* for small orchestra, composed about 1760. The *Divertimenti*, although the material of the outer movements is slight, show the energy of his thought in the continuity and balance of swiftly moving phrases. An occasional sudden irregularity gives great impetus to the music. These are five-movement works – *Allegro*, *Minuet*, *Andante*, *Minuet* and *Finale*. The minuets are stately and simple. Already there are signs of symphonic mastery. This is still more true of the *Scherzandi*, where Haydn begins to show a very strong Viennese influence (particularly that of Wagenseil) in decoration which is so absorbed into the structure of the music that it is no longer purely ornamental. This is a marked characteristic of Wagenseil and others of the

'Proper teachers I have never had. I always started right away with the practical side, first in singing and in playing instruments, later in composition. I listened more than I studied but I heard the finest music in all forms that was to be heard in my time, and of this there was much in Vienna.'

Viennese school, as it is not of the Mannheim composers or of C. P. E. Bach, a major fault of whose symphonies and sonatas is that their ornamentation remains superficial, however beautiful.

In 1759 or 1760, Haydn wrote the first of the 104 symphonies, in D major. It is a three-movement work, without the minuet of the *Divertimenti*, and begins in a way no self-respecting Mannheim composer would have considered – quietly, with a crescendo following. This kind of beginning, however, can be found in Wagenseil and Starzer. Not only is there no *coup d'archet*, but the first phrase is an indivisible nine bars growing from *piano* to *forte*. In a long and beautifully balanced passage of tonal transition, where the proportions are already characteristically inevitable but unexpected, he makes richer use of some of Carl Stamitz's subtleties, so that his second subject appears to arise in mid-phrase, with the harmony suspended. The slow movement is, as often at this period, a simple piece for strings alone with a touch of the relaxed contrapuntal motion of earlier times, though it never became really contrapuntal. The finale is the most startling of the three movements. Here, at the beginning, we are shown in surprising fashion just how much speed can be generated by ambivalent phrasing:

14

It will easily be appreciated that this curious phrase can be heard in two ways – either that marked *a* or that marked *b*. What follows shows that either makes admirable sense, although, of course, the sense of the rest is different according to which view one takes. It is followed by a fourteen-bar phrase, which leads to the second subject in a four-bar phrase extended to six, answered by another six. It is this, far more than any rapidity of notes, which gives this movement its impetus and makes it large, although it takes but a few minutes to play. This first symphony is, moreover, remarkable for the fullness of sound extracted from the normal orchestra of the time – oboes and horns with strings.

The second symphony (1760) begins to show something that distinguishes Haydn from most of his fellows – what he does in one work is no indication of his methods in another. The main theme is as simple as it can be:

15 

but with its quiet counter-statement some bars later he sets his seal on it with a treatment that is solemn and frivolous at the same time:

16 

At the end of Ex. 16 I have marked a group of notes with which Haydn shows himself the complete romantic, in a passage designed to settle fully his second key of G major. This could well appear in a work of the early nineteenth century:

17 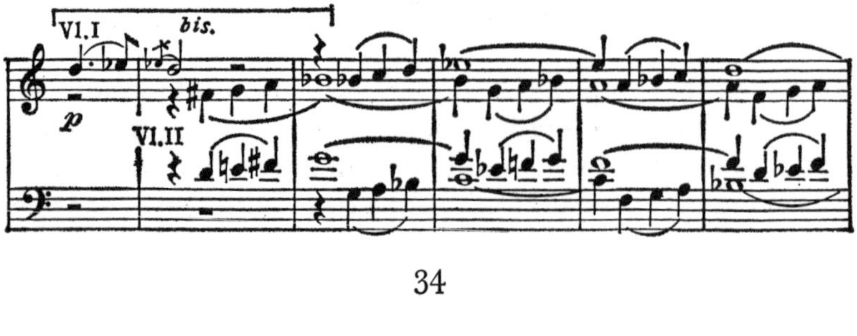

Much of this movement is in a simple two-part counterpoint; one of the means by which Haydn brought new life into the symphony (and sonata style in general) was counterpoint – of the right kind, at the right places. We have already seen him doing this in a later piano sonata movement and it is, of course, what Mozart does in the finale of his last symphony. It is a new kind of counterpoint, invented for this specific purpose, and how different Haydn's fugal counterpoint was can be heard in any of the early string quartets which possess a fugue – the magnificent one in the F minor of Op. 20 for instance.

This kind of counterpoint can be heard very clearly with the beginning of the third symphony of 1761:

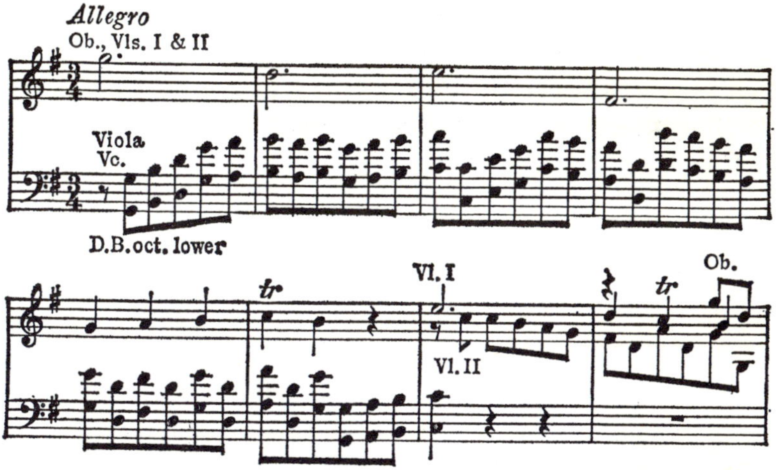

18

Ex. 18 shows the simply moving two-part writing of the opening (which returns intensified in the development) and the dropping of this texture as the music approaches the change of key. As if to emphasize this more strongly the finale is a mixture of full-blooded fugal writing and homophonic sonata 'business'; it must be the earliest attempt to reconcile these two intractable styles:

19 

One new prospect after another opens to our ears as these works succeed each other. There is Haydn's ever-new way of designing a highly personal theme from some pregnant harmonic process, as in No. 5 in A:

20 

and his use of the solo violin and 'cello in Nos. 6 and 7, *Le Matin* and *Le Midi*, and No. 8, *Le Soir* (or *La Tempesta*), so-called because of the stormy finale. There is the ending of No. 9 with a minuet, a movement which creeps in and out of these early symphonies, sometimes as the third of a four-movement work, sometimes as finale of a three-movement work. There is the beginning of No. 11 with an *Adagio* followed by a normal *Allegro*, where this order of movements sounds right and the usual order would not. And there is the infinite variety of ways in which Haydn shows us that the establishing of a second key to replace the lost original one is for him the main division of his exposition, and that where he often has no definite second main theme he may end the exposition with a beautiful new soaring version of the first theme, having occupied most of his 'second subject' space with harmonic assertions and explorations of the new key. There is also the great variety of his slow movements, from his inimitably calm, grave kind of beauty to the quizzical semi-ponderous opening *Adagio* of No. 22, with its two cors anglais.

No life the length of Haydn's can be lived at the same temperature throughout. External events will have their effect, and we know that in the main Haydn's life appeared reasonably happy. We know, too, that he had an early unhappy marriage to live down, and perhaps, since he had an active conscience, some trouble to decide just where he had failed in this partnership. The one or two symphonies from Haydn's mid-career which have been at all frequently played, such as the 'Farewell' (No. 45), are usually far too much associated with irrelevant stories. Everybody has probably read the story connected with this symphony; it is a good joke. But the music accompanying this joke is not funny; it is contemplative, with a gravely disquieting sort of happiness; a kind of contradiction common in the work of this only apparently simple man.

Haydn's music shows him to have led an intense inner or spiritual life; he was a modest man, and it is natural that he should have hidden this as much as possible from the world under a guise of cheerfulness and high spirits; but his music could not hide it. There is in all the symphonic writing (and in a number of the string quartets) of the late seventeen-sixties, seventies and early eighties a dark, brooding, even fiercely burning passion. No one can know what spiritual experiences this music commemorates, but it is understandable that the prevailing happiness of the later symphonies is an expression of the gratitude Haydn felt towards his Maker for the life of grace he had been permitted to live.

It would require a book to describe the great range of expression in the darker middle-period works. I have selected one as an example, the F minor, usually known as No. 49. Research has shown that this work should more properly be placed after No. 35. The title by which it is generally known, *La Passione*, is not Haydn's and appears only in the early nineteenth century; it is truer to the facts than most such titles. Certainly it would not be out of place to regard the symphony as a meditation on the Passion of Christ, when it would carry the same conviction as any piece by J. S. Bach inspired by the same subject. All four movements are in a sombre F minor, reinforced by equally sombre instrumentation, for oboes, horns, strings, and (Mr Robbins Landon thinks) continuo also, with a bassoon playing at certain specified places. It is true that Haydn does frequently indicate a passage for solo bassoon where there is no bassoon indicated in the score; this means that its presence as a continuo instrument was taken for granted. Although he indicates no continuo part in his scores, to the end of his life he directed performances of his symphonies from the (pianoforte?) keyboard and often improvised a continuo part. How much this was the result of habit and how much a realized necessity, it is difficult to say; there are certainly cases where it adds a very definite aura to the music; there are equally cases where I, at least, feel its employment to be an intrusion not warranted by the music. There are many early slow movements which gain in intensity without it. And there is the fact that Monn, who came earlier to the style, nearly always indicates continuo in his score. There are far too many rival theories today for us to be absolutely sure about this point, and it is a musical and historical sense which must guide us in the end.

The present case is one which can be argued both ways. As with so much of Haydn's music, the work often resolves into two-part writing, and some would argue that without the continuo the music is very bare. True – and in some cases this would be a sound argument. Here, I am not so sure; this bareness of sound is so much in the character of the work, it is maintained so consistently and falls in so much with the persistence of the main tonality, that there is almost a sense of one huge organic movement rather than four shorter ones. I cannot help feeling that the absence of the continuo is justified in the result.

The symphony divides itself into two halves, the opening *Adagio* from which the succeeding *Allegro di molto* is a reaction and the Minuet and final *Presto*, likewise a complementary pair of movements. The unity of mood and a strong emotional and thematic continuity between first and third, second and fourth movements is a new resource in symphonic writing, obviously the outcome of a deep personal emotion; it is impossible to relate this music to the stiff, formal, polite small talk of the period, to anything as superficial as 'rococo' or, at the other extreme, to the *Sturm und Drang* movement of Lenz and Klinger so popular at the time. The thematic unity across the movements can be heard quite simply:

(a)

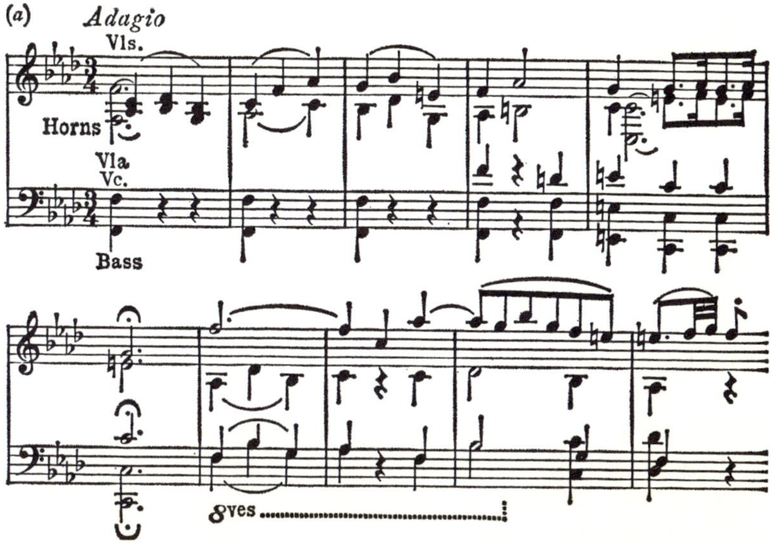

21

The minuet theme, it will be noticed, is a combination of both the themes shown in Ex. 21 (*a*). Here is an example of the bare string writing in the *Adagio*:

22

It is from the transition passage, which I feel is characteristic and would be spoilt by a continuo part. The 'second subject' is merely a tonic and dominant assertion of A flat, with telling sustained 'continuo' chords on the wind, followed by the semiquaver rhythm of Ex. 22, now in the form of broken chords. When Haydn does introduce a new theme it is, as so often in his later works, a cadence theme, the first of two. The second of these has a broken rhythm which is infinitely moving.

The development presents most of the material of the first group from bar seven of Ex. 21 (*a*) onwards, but with a more intense sound derived from the major key, and at times a more intense scoring. For this reason the restatement of this group is much curtailed.

I feel that the second movement should break out without a pause, its mood generated by the first:

23

Ex. 23 shows also the cross-currents between this movement and the finale. Haydn's main theme in the *Allegro di molto* makes use of the dropping diminished seventh which is almost to be taken for granted in music in this mood, from Bach onwards; and the sharp bitterness of the mood is emphasized by the biting sound of the oboes in unison with the violas, 'cellos and basses, another effect which would be ruined by a keyboard continuo.

Haydn's way of approaching his second key of A flat is an extension of Sammartini's method, described earlier. His first subject rushes quickly to the home dominant, and a quieter but no less swift theme plunges immediately on to A flat. This could be taken as Haydn's 'second subject', thematically, but he shows that he does not consider that he has established A flat as an independent tonality. It is really the start of a large transition passage in which Sammartini's method of emphasizing the dominant of the new key *within* his 'second subject' is developed with a continuity of which the Italian composer never dreamed. When the second key is eventually established, it is with a new theme based (accidentally or intentionally) on bar 10 of Ex. 21 (*a*), and the music receives a great forward thrust from the phrasing (3 + 3 + 9 bars).

This brings about the end of the exposition. In its development and recapitulation this *Allegro* reflects the process of the first movement – first group and transition material intensified in the development, with a curtailed restatement as a result. The two movements are one conception. Never before had the Minuet, normally a light relief, been so transformed; it is perhaps the most sombre movement in the work, although there is one akin to it in the string quartet in the same key, Op. 20. Most telling is Haydn's way of allowing the oboes to play many phrases in unison with the first violins, in each case followed by an echo for strings alone; the trio, in the major, relieves this pattern, and the oboes are given some independence for a few bars, on almost the only occasion in the work. This is another argument against the use of the keyboard continuo in this symphony; the wind instruments are supplying an effective continuo throughout.

With the finale there is again a sense of reaction from the mood of the preceding piece (see Ex. 23(*b*)). Its transition is effected in precisely the same way as in the second movement, and one of the most striking momentary effects in the work is a triplet of crotchets at the end of the first subject, just before the transition passage begins:

24

I have heard this debased to the rhythm of two quavers and a crotchet, and once, (even worse) to that of a crotchet and two quavers. The development again makes much of the material of the first subject, starting from A flat. But there are differences here; the opening of the movement, Ex. 23(*b*), is joined to the latter part of the transition passage to form a new theme, whilst a passage in C minor (leading to the recapitulation) gives the opening figure again with a connecting link of crotchets which makes an unbelievable difference to the recapitulation of this idea a few bars later, when the crotchets are dropped.

One of the most forceful elements in this symphony, having great importance in Haydn's development, is the organization of the two quick movements, with their avoidance of artificial divisions and, often, even points of rest. Each sweeps on in one great span. Although it was some time before Haydn attained quite this degree of continuity again, the later symphonies are the fulfilment of his aim, and belie the conventional idea of the cut-and-dried, velvet-jacket-and-lace-cuffed 'Father of the Symphony'. The F minor symphony is a landmark in Haydn's work; it marks both the end of one phase and the beginning of another.

One of his notable additions in the late symphonies is the slow introduction. Haydn and the slow introduction are almost as inseparable in most people's minds as Haydn and the symphony or Haydn and the string quartet – or, for that matter, as Schubert and song. In fact, of the first eighty-three symphonies only thirteen have slow introductions. From No. 84 onward only three are without: Nos. 87, 89 and 95. It has been suggested that the introduction is one of the features taken over from the *French Overture;* the objection to this is that it was rather late in the day for this to happen by the time Haydn wrote the first of these works to have such an opening, No. 6, about 1761. This was not a feature of the early Viennese or the later Mannheim composers in their symphonic writing, although composers like Monn continued to use the normal introduction in the *French Overture.*

Haydn's introductions have none of the characteristics traditional in those of the *French Overture.* He varies them greatly. At times they simply introduce, as briefly as possible; sometimes they seem about to grow into movements themselves; but the composer never loses sight of the fact that an introduction is not the main matter. It can be solemn, like the great portals through which we enter the first movement of Mozart's last E flat symphony, for solemnity's sake, or (never in Mozart) to point a great joke. Or it can be mysterious. Whatever it expresses, its character is always related to what follows. It would be impossible to transfer the introduction of one Haydn symphony to another.

It is always difficult to select from such a body of work as Haydn's, but as representative of his last period of symphonic writing I have chosen No. 97 in C, composed for his first visit to London. The first movement is, even for Haydn, of astonishing energy, and the nature of the slow introduction is determined by this, expressing, as it does, the calm strength necessary to sustained athletic energy:

25

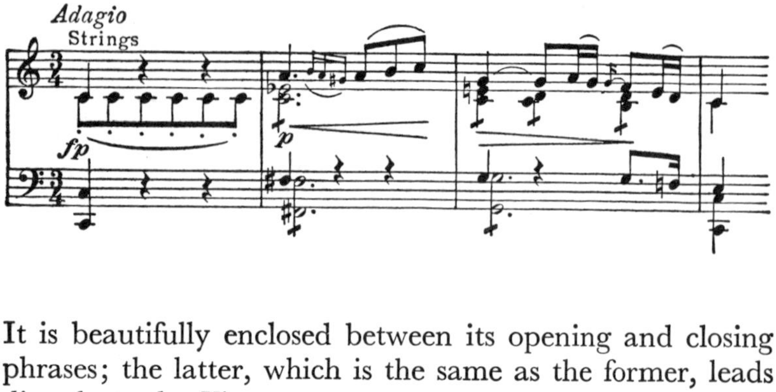

It is beautifully enclosed between its opening and closing phrases; the latter, which is the same as the former, leads directly to the *Vivace*:

26

The main movement begins in a way unusual in the later symphonies, with a theme which is purely a theme and nothing else. From his early days when the theme was more often treated by Haydn and others as a necessary prop to the key, he has developed the freedom to design themes of strong melodic character. Such a theme as we have in Ex. 26 is by now a rarity. Themes of this kind evoke comparison with Beethoven, who is notably given to the blunt assertion of them. Very well; there is a connexion between this movement and Beethoven, and we will see exactly how far it goes. For the moment let us follow the course of the exposition; Ex. 26 has a sequel:

27

expanded for seventeen bars. This sets off the transition passage,

28

beginning with a diminished version of Ex. 26(*a*), accompanied by the quavers of Ex. 26(*c*) and continuing with Haydn's own brand of fanfare figures until the music settles down into the 'second subject'. Haydn again departs from his more customary practice by having a new theme, actually a tune, to accompany the second key of G major; it is in waltz rhythm, with the bassoon obligingly completing the rhythm in each bar:

29

This, followed by energetic triplets and explosive off-the-beat accents, takes us cadentially to the end of the exposition, ending in restrained quiet. Now we can perceive Haydn's proportions in this exposition and so begin to see what he considers important. Remembering the thematic material and the one tune, we have a first group occupying twenty-six bars, a transition passage of thirty-six bars and a 'second subject' of thirty-two. Haydn has made his emphasis abundantly clear. This is one point in which he differs from Beethoven, who, while putting as much accent on key, never does so in these proportions; and such proportions are more the rule than the exception with Haydn.

With the development he shifts the music up a sixth to E flat, delivers the main theme shortened by one bar, and continues with Ex. 26(*c*). Haydn does not concentrate upon this quaver figure, as Beethoven would surely have done. Instead comes a characteristic piece of continuous atonality (or, if the reader prefers, a passage in which no key is established), a kind of music increasingly common in a Haydn *Allegro* at this stage. Moving (in both senses of the word) is an intricate piece of three-part woodwind counterpoint:

30

reinforced with echoes of Ex. 26(*b*). Out of this comes one of the most imaginative of Haydn's instrumental passages, the falling arpeggio of the main theme striding in sequences to the recapitulation, with the bassoons doubling the first violins in a gradual anticipation of Ex. 26:

31

The original expansion associated with Ex. 27 is curtailed, with one of Haydn's grandest harmonic surprises bringing the 'second subject' into C major, and the quiet conclusion to the exposition is expanded to form a coda of a size one might find in early Beethoven. Parts of the transition omitted in the recapitulation are now included, and the end is energetic.

If this movement has been absorbed, the other three become easier to digest, and it will be seen that there is more in them than individual beauties. There are many subtleties in the beautiful and simple variations of the *Adagio*, such as the telling refrain to the second part of the theme:

32

or the way in which the wind reinforcement of the phrase-ends is maintained unchanged in most of the variations. Subtle also are the Minuetto's expanding cadence-rhythms:

33

Growing out of the main theme, these give the piece an organic unity strong even for Haydn in such a movement. The trio is a frankly simple and straightforward tune, with a quietly chuckling accompaniment:

34
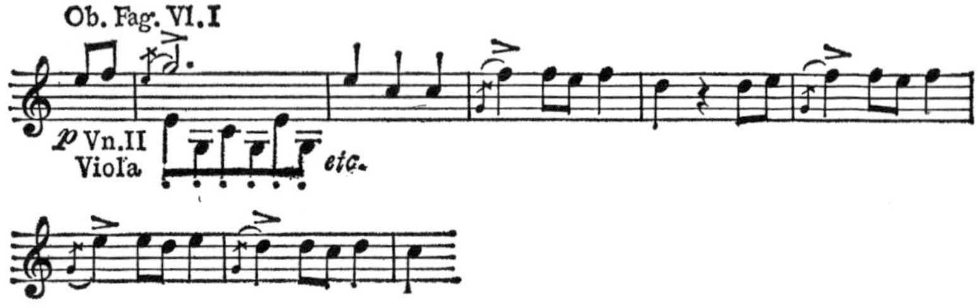

The feline grace and speed of the finale, a magnificent rondo, embody Haydn's humour as he sums up the whole work in a great burst of energy. One extract will show the vitality of the music, after the characteristic four- and two-bar phrases of the main theme:

35

Haydn was the first supreme master to make the symphony and sonata the foremost vehicle of his thought. He retained pre-eminence until Beethoven arrived, for Mozart, great as are some of his symphonies, rarely gave this medium the same attention as Haydn. No later symphonists have improved upon Haydn's position; no one can do more than equal him; even Beethoven, in some ways, scarcely surpassed him. One thing alone would mark him out – the variety and range of his thought in this mass of works. Each symphony is a world in itself, even the smallest and earliest. There is no repetition. In this sense, and it is a deep one, he *is* the 'Father of the Symphony'. He first perceived and expressed the heights and depths this kind of music could attain.

2

WOLFGANG AMADEUS MOZART

(1756–91)

Hans Keller

*

Two things must be realized at the outset of any musical (as distinct from a merely musicological) examination of Mozart the symphonist. In the first place, it is unfortunately necessary to remind ourselves that Mozart would have laughed at the very idea of an extended chapter on his symphonies. 'What is there to write about?' he would have asked in genuine amazement. 'Many of the earlier ones are too bad, and most of the later ones too good to stand in need of explanation.' Ignorant as he was of such terms as 'first subject' and 'second subject', he would not have been aware that a professional analyst could in fact write about anything without meaning anything or even being supposed to mean anything; that there would ever come a time when there was a readier market for the meaningless than for the meaningful. In art, the meaningless *par excellence*, which is always seemingly (read: historically) meaningful, is that, and that alone, which several works and several composers have in common – first and second subjects, for instance. So long as such common characteristics, which I call form, are not related to a work's individual thought, which I call structure, the information conveyed remains essentially statistical and inartistic. The meaning of a musical work consists, in fact, in the tension between background form and foreground structure: the foreground is composed *against* the background, which is the basis of communication between the composer on the one hand and the player and listener on the other.

The question arises why, nowadays, there is such a demand for formal description. The answer is – insecurity, an insecurity which is only partly of our time, inasmuch as it is part of our general artistic crisis, in the course of which we have arrived at a state of bewilderment which makes us welcome anything that replaces rather than explains our waning musical experiences, until we are glad to be told what we ought to feel. The other part of our unobtrusive insecurity about the ever-present past, about 'classical music' as we call it, is but the tail-end of a process that started in the romantic era, when the historical distance between the player or listener and his most beloved music began to make itself felt, a process of semi-conscious estrangement that has increased with that distance until we find people who like, say, Mozart 'best' for no more positive reason than that there is nobody else to like best anyway. 'It is strange', says Einstein, 'how easily the world has accepted [the great G minor symphony] and has even been able to think of it as a document of "Grecian lightness and grace" [Schumann's famous description] – a characterization that could apply at best only to the divine tranquillity of the Andante or to the trio of the Minuet, otherwise so heroically tragic.' 'Strange' is the word; the estrangement had started in Schumann, although he was as yet entirely unconscious of it.* And then the supreme paradox of which Einstein seems clearly aware: the ready acceptance of music not because it is understood, but because it isn't. The idiom sounds familiar (apart, that is, from the development section of the G minor's finale, of which more later), so what more do we want? Do we really want to get down to the clear substance, when pleasantly vague feelings are so readily aroused on the surface? In a way, therefore, 'distanced' classical music is worse off than the most difficult contemporary music – in that it is wide open to misunderstanding,

* But we should beware of entirely disregarding a genius's positive reaction to a masterpiece, however distorted his outlook may have been. The element of Grecian lightness and grace reappears quite naturally, together with the tragedy, where we should have least expected it – in Furtwängler's interpretation of the work, which has been preserved on gramophone records.

whereas modern music, at its worst, is merely closed to understanding.

In the circumstances, the duty of the musician writer on classical music is plain. He must sell the dummy to history herself: feigning to meet her requirements for formal description, he must re-establish contact with musical substance, i.e. his own musical experience as well as that which lies at least latent in the player and listener. The only way to remove insecurity is to show that it is unnecessary, and the only way to talk about music is to talk about one's addressee's own feelings, actual or potential; everything else is sham. The analyst is not there to instruct, but to clarify. You cannot instruct a stone about the purpose of life; but from a dying man you can learn to clarify, both for yourself and indeed for him, the meaning of death. A critic who is not as much in touch with his audience as with the music – a critic who is not basically a teacher – is a charlatan.

The second thing we have to realize before we actually concern ourselves with Mozart's symphonies is that if he had been forbidden to write any of them, he would not, musically, have been maimed. This is not to say, of course, that he did not write some of the greatest symphonies ever (the last four to be exact), or that he was not intrinsically a symphonist. But an essential – and perhaps the most original – part of his symphonism (by symphonism I simply mean large-scale integration of contrasts) he poured into his piano concertos and his string quintets: it seems to me that his predilection was for two contrasting 'tune-carriers' – piano and orchestra in the concerto, first violin and first viola in the quintet. (Nor must we forget that he liked to play himself what he composed – on the piano or on the viola, as distinct from the violin, of which he was not so fond, perhaps owing to his ambivalent attitude towards his father.) The significant fact to note is that there is no important medium in which Mozart wrote so many unimportant works as in the symphonic field proper. No more than one-fifth of his symphonic output dates from the years of his greatest maturity, his last twelve years;* and in the last three years he did not write any symphonies at all. It is a little facile to say that after the finale of the 'Jupiter', no further symphonic thought was possible or necessary for him; if Beethoven had ceased symphonic writing after the 'Eroica', an even stronger case could have been made for that work being an end, a symphonic testament. By all means, let us be wise after the event, so long as we are sure that we are being wise.

It is indeed the simultaneous familiarity and distance of Mozart's language that is responsible for the recurrent, unobtrusive over-estimation of his early symphonies, which have little to do with the symphonic concept as we know it – not, in the first place, from Beethoven, but from Haydn (whose string quartets, incidentally, are even more symphonic than his symphonies), and from the later Mozart himself. After all, these early symphonies were written by a child, who has wrongly been credited with prematurely developed genius (which we do find in Schubert and, above all, in Mendelssohn). What was so incredibly precocious about Mozart was his talent, and it was the very facility springing from it that inhibited the growth of his genius: his true personality emerged later than Beethoven's, who was not blessed and cursed with that credit account that is a genius's initial dexterity.

*

Why K. 16† in E flat major (1764/65) should ever be performed, is incomprehensible; yet we see earnest musicians like Hermann Scherchen plough through this 'symphony' with what can only be assumed to be interest. There are, of course, better symphonies by some of Mozart's lesser known contemporaries and forerunners, above all by

* Nevertheless, in an immodest effort to make this analytic survey more constantly helpful than any of its predecessors, every single symphony will be included in it.

† Throughout this chapter, works are discussed in order of the 'Little Köchel' numbers, a procedure which makes reference much easier than the usual, strictly chronological order. The 'Little Köchel' has been chosen as the most readily and practically available source.

his unintentional teacher, Johann Christian Bach; apart from which we know, in any case, too much music rather than too little – or rather we hear too much and know too little. Anybody who has a legitimate reason for wanting to find out about the nine-year-old Mozart's symphonic 'style' can have a look at the scores which, we may assure him, are not difficult to read. If the three movements of K. 16 ran into each other, they could be a primitive – albeit too slight – Italian Overture. For what they contain, they take up an inordinate amount of time – about a quarter of an hour with repeats, of which there are plenty.

K.16a and 16b (both dating from 1765) are not extant in either autograph or print; the chances are that we haven't missed much. K.17 in B flat major is more than simple-minded, its authorship in fact highly dubious, and K.18 in E flat major has turned out to be by K. F. Abel, though it is in Mozart's hand: the boy copied out the whole of the symphony. This method of study, normal at the time, has now been almost entirely forgotten. The reason, it seems to me, is not merely the loss of a common language, but the associated, partial loss of hearing-while-you-write: nowadays, few composers are left for whom writing something means hearing it through, every note and every chord of it.

K. 19 in D major, was written in the spring of 1765 – in London, as was K. 16. Jens Peter Larsen calls it 'an astonishing advance' on that first symphony; 'one would hardly judge that its composer was still a child. It is much more "grown up", and displays a mastery of melodic and formal resources that can be taken as indicating a certain "maturity".' I must leave it to the reader to make all this out, in particular the meaning of the word 'mastery' in such a context. For my own part, I find the earlier symphony more touching, though only retrospectively: the E flat work's latently expressive *Andante* in C minor, coming from a composer who, otherwise, never came to show a particular liking for slow movements in the minor mode, offers a distinct glimpse into the future, for E flat was to remain the only major key whose relative minor Mozart repeatedly reverted to for slow movements. Curiously enough, the Abel symphony (K. 18) shows the same key relationships; in fact, on internal evidence, I would suggest that Mozart copied it before, not after, he wrote his first symphony (K. 16).

With K. 19, then, we leave what would be known as Mozart's 'London' symphonies if they were a little more interesting. Although K. 22 in B flat likewise dates from 1765, it was probably written in The Hague. (In between come K. 19a in F major and 19b in C major, both from the same year and both lost.) The *buffo* style, which has been present from K. 16, continues; in fact, the last movement (*Allegro molto*) looks far, though not very substantially, ahead to a comic opera, in that its first four bars are taken up literally at the beginning of the G major section (*allegro*) in the second-act finale of *Figaro*. How meaningful the phrase, fairly meaningless in itself, has there become! In the early symphony, it stresses itself, the continuation being no more than an extension. In *Figaro*, apart from forming an ironically liberating contrast to the preceding B flat major section, it stresses the contrasting entry of Figaro ('*Signore, di fuori son già i suonatori*'), to which, as an antecedent,* it yields. In short, in the symphony the phrase is composed *with*, while in *Figaro* it is composed *against*, serving in fact as one of those backgrounds against which significant composition establishes its tension. The time-honoured argument whether or not a tag is worse as material for extended composition than a characteristic thought is irrelevant without reference to the function of the motif, phrase, or theme in question. As a foreground – see K. 22 – it is worse, as a background – see *Figaro*, K. 492 – it is as good.

The three Vienna symphonies, K. 43 in F major (1767), K. 45 in G major, arranged, after a few months, to serve as overture to *La Finta Semplice*, and K. 48 in D major (both 1768) replace some of the impact of Abel's work, as well

* The terms 'antecedent' and 'consequent' are as yet more frequently used in American than English musicography. If I may give an informal explanation, an antecedent is that to which the consequent, the answering phrase, responds.

as the deeper influence of J. C. Bach, by that of the Viennese tradition. When we think of all those bores, the Wagenseils,* Hofmanns and Gassmanns, who evidently left a deep impression on Mozart's growing creativity, we get a measure of the distance between even the most spontaneously sensitive classicists among us and the actual classical style with which, they think, they are in immediate touch: those composers weren't bores at the time. Of course, there is always 'historical listening', which means talking oneself into an interest one does not feel, but if musical understanding stands and falls with spontaneous experience, such artificial measures hide a symptom under the cloak of a cure.

Under this Viennese influence, the three-movement work tends to become a four-movement one, the dance movement (minuet) being added in third place. The three-movement K. 45a in G major, and the four-movement K. 45b in B flat major (both from 1768 and neither generally available) may well have been written in Vienna too. Later on, when we come to discuss the 'Prague' symphony, K. 504, we shall see that while the three-movement scheme is, in itself, more primitive than the four-movement one, it gave rise, as a formal background, to one of the most sublimely developed structures in the history of symphonic discovery.

K. 66c in D major, K. 66d in B flat major, and K. 66e in the same key (all dating from 1769), are not extant. K. 73 in C major (1771) is in four movements, but in between, chronologically, comes amongst others K. 74 in G major, one of the 'Italian' symphonies (as we might call them) which, associated as they are with Mozart's two journeys to Italy, show renewed Italian influence. In fact, the work reverts once again to the three-movement layout, with the first movement running into the *Andante* in the manner of an Italian overture. It would appear that at the beginning and towards what might prove the coming end of the history of the symphony as we know it, its formal scheme is equally flexible. It may indeed be that we here have a piece of truth about the history of form and technique in general: during the

* Harold Truscott has a different opinion of Wagenseil. (Ed.)

periods of initial integration and eventual disintegration, reintegration, or re-orientation, relative freedom obtains; whereas in the middle, at the classical stage in the development of any particular means of composition, we encounter the kind of strictness that makes the later publication of text-books possible. In the beginning, in other words, the formal principles have not yet crystallized, whereas in the end they would calcify if they were not resolved. This type of evolution can even be observed in the history of a single composer's work – if his is a strongly developing creative character, and if he survives long enough for his work to pass through those three periods of which Beethoven's output is, of course, the outstanding example.

K. 74g in B flat major was written between Mozart's two Italian journeys in the early summer of 1771; it has a minuet, with a trio that is a 'quartet' – for strings alone. K. 75 in F major dates from the same period and evinces a remarkable predominance of triple time; only the slow movement (*Andantino*), which now comes third, escapes this metrical scheme: it is in duple time. Historical influences apart, this 'triple time' approach will usually be found either in small composers, or in great composers in their early stages of uncertainty. The reason is that triple rhythms take over part of the task of characteristic invention: other things being equal, an undistinguished idea in triple time will sound better defined, more 'individual', more rhythmic in fact, than one in duple or common time, which forms a more neutral background. We shall never, in the mature Mozart, encounter K. 75's type of metrical scheme again.

K. 76 in F major is a much earlier work; it was written in Vienna in the autumn of 1767. While it is a four-movement symphony as it stands, Einstein suspects that the minuet may have been written 'into it' at a later stage – a view supported by the fact that it is the only movement that holds any interest.

K. 81 and K. 84, both in D major and both written in 1770 in Italy, belong unmistakably to the Italian group: they are both without minuets. But K. 95, again in D and also an Italian symphony from the same year, has four movements, though the minuet seems another of those later additions to a typically Italian form whose first movement runs into the second.

K. 96 in C major, with trumpet and drums, was written in the following year in Milan. The four movements are an interesting example of what I call homotonality, the slow movement being in the tonic minor. A notable fact which is never remarked upon is that although Mozart, as opposed to Haydn, tended to work within narrow tonal frameworks, he did not carry the homotonal approach very far into his maturity (though the wind *Sinfonia Concertante* and the 'Turkish' Piano Sonata, both written in 1778, are both homotonal*) whereas Haydn did: some of the older master's greatest string quartets, in particular, adhere to a single tonality.

K. 97 in D major (Rome, April 1770) is a normal four-movement work; Einstein again submits the possibility that the minuet is an afterthought. The scoring includes trumpets and drums; the *Andante* (in second place) and the trio are for strings alone. The next Köchel number, K. 98 in F (1771), is dubious. So long as we can doubt the authenticity of a work, we know that we have not yet arrived at the real Mozart. The case of certain passages or details in the *Requiem*, which sound authentic but about which there may yet be some musicological doubt, is different: if at any future stage it should turn out that Süssmayr composed some genuine Mozart, we shall not be musically perturbed, for it is conceivable that he wrote such music as a result of his master's hypnotic influence; after all, Süssmayr's creative mind was, as it were, Mozart's creation.

K. 110 which the 'Little Köchel' (p. 74) wrongly describes as being in C major, whereas it is in fact in G, was written in Salzburg in 1771. Back under the Austrian influence, the four-movement work contains, according to Larsen, 'little that points to the future' – which is true enough so long as

* See p. 75.

we talk in terms of foreground, and foreground alone. But the *Allegro* in 3/4 time that opens that symphony emerges as a distinct background, however remote, of the principal structure of the late E flat symphony, K. 543. I mean the opening of the first movement's *Allegro* with its 'time-beating' first beat which, in melodic reality, introduces a phrase and theme that starts on the second beat, i.e. with an upbeat motif: it is this very device which, in embryo, can be observed at the beginning of K. 110, with its wide leaps from the first beat tonics to the conjunct motion starting on the second beats of the successive phrase. Again, the rondo theme of the final *Allegro* shows an interest in the rhythmic structure of the gavotte which was to remain with Mozart all his life – as witness not only such obvious examples as the C major piano concerto, K. 503, with its re-use, in the finale, of the *Idomeneo* gavotte, but various *slow* movements in which the gavotte rhythm operates as a suppressed, yet palpable background: in the *Kleine Nachtmusik*, K. 525, the sole E major piano trio, K. 542, and the last (E flat major) string quintet K. 614. In all these instances, Mozart exploits the gavotte's natural ambiguity or rivalry between the first and third beats – which in the youthful K. 110, he evades by casting his theme in a 2/4 (as distinct from a 4/4) pattern: but the very evasion shows his growing awareness of what was to become an intensely fruitful problem for him.

K. 112 in F major is a four-movement symphony written in Milan in 1771; it is one of the works where, according to G. de Saint-Foix, 'the first symptoms of what is to become in the course of time the real symphonic art of Mozart make their appearance in striking fashion', whereas according to Larsen it indicates once more 'that his symphonic development at this time had not reached any new creative phase': choose as you like, remembering that while these authorities cannot both be right, there is no logical complication involved if we assume that they are both wrong. Where we have to follow Larsen, I think, is in his comparative evaluation of the movements: only the first is of any real musical interest. He calls it 'Haydnesque', and while we can feel what he means, we ought not to leave our feeling unanalysed: the opening phrase of the 3/4 movement is identical with the opening phrase of Haydn's string quartet in D major, Op. 71, No. 2 – which was written twenty-two years later! The trio of this work, incidentally, is once again for string quartet.

K. 114, which was composed at the end of 1771 in Salzburg, does, however, mark a new period in Mozart's symphonic output. The key itself is significant: the work is the first of Mozart's three A major symphonies, the others being K. 134 and *the* K. 201. From now on, Mozart's symphonies were, in fact, to evince greater tonal variety as between works. And in the thematic dimension the tendency towards extreme contrast began to assert itself within the works. For while Mozart never showed any intention of emulating Haydn's tonal and harmonic adventurousness, he now started to show his true, polythematic character – an adventurousness of melodic invention, variegation and integration which Haydn, always a monothematicist at heart, would never have dreamt of emulating. An important aesthetic fact emerges from this comparison between the two geniuses, opposite creative characters as they were in most respects: true greatness tends to confine its newness and complexity to one dimension – thematic structure in the case of Mozart, harmonic structure in the case of Haydn. The other dimensions remain simpler and thus serve comprehension of the new. It is only where nothing 'simple', nothing old remains that the new is destined to remain incomprehensible. Some of the thematic contrasts in this immature Mozart symphony, K. 114, are greater than the corresponding contrasts in the maturest Haydn who, instead, concentrated on harmonic contrasts.

K. 124 in G major (Salzburg, February 1772) and K. 128 in C major (Salzburg, May 1772) both have their first movements in 3/4 time, but although the Austro-German influence continues, the latter work follows the Italian overture's layout, as does K. 129 in G major, written at the same time, whose finale starts like the opening movement of

Mozart's last piano sonata, K. 576 in D major: as in the case of the gavotte, we here encounter a typical last-movement pattern which, in Mozart's maturity, was to be transplanted to an opposite movement. The expressive potential of that first movement was thus heightened through the tension between the audible foreground of the music and the 'finale' feeling lurking in the definable background.

K. 130 in F major, composed at the same time (though probably at the end of May), returns to the four-movement scheme. The symphony shows at least two new aspects. For one thing, there is a conscious, perhaps even conscientious feeling of unity between the movements, however unsubtle as yet: all four movements start with variations of the same basic motif. And at the same time, Mozart is already careful to confine close thematic relation and resemblance to movements with contrasting keys – hear the beginnings of the middle movements, in B flat and F respectively. For another thing, the *Andantino* (*grazioso*) is, for the first time, in triple metre; Haydn, incidentally, introduced the innovation in the same year.

K. 132 in E flat major and K. 133 in D major (Salzburg, July 1772) are remarkable for their *Andante*s in the dominant, as distinct from the customary subdominant. Other things being equal, the choice of the dominant invariably means greater liveliness or even tension than is possible in the reposeful subdominant, whose establishment as a tonic, for the duration of a movement, inevitably creates a feeling of relaxation. The trio of K. 132 is of the 'quartet' kind we have encountered before; but a striking anticipation of the more remote future is to be found at the beginning of the work: it forms a patent source of the corresponding theme in the piano concerto K. 482, which is in the same key.

K. 134, the second A major symphony, was written in the following month; in fact, the key – Mozart's principal 'love' key in the operas – seems to have been in the air: about two months later, he went to the A major dominant again (rather than to the subdominant) for his slow movement in the D major quartet, K. 155, as he had done in K. 133.

(Apart from a fragment he wrote only two string quartets in A major: one of them, K. 464, is his greatest symphonic structure in that key, a work which Beethoven profoundly admired.) 'A new type of singing character' (Einstein) is evident in this A major symphony. The key evidently meant a great deal to the composer of Ferrando's love aria, and the love duet in *Così fan tutte*, not to speak of such other operatic pieces as *Là ci darem la mano* or the trio, *Ah taci, ingiusto core!* in *Don Giovanni*. It is surprising that Einstein, in his chapter on *Mozart's Choice of Keys*, says so little about this one; 'A major, the boisterous' is about all he gives by way of characterization.

The first two movements of K. 161-K. 163 (or K. 141a according to Einstein's revised Köchel) were written in March, 1772 in Salzburg, while the finale was added towards the end of that year in Milan: hence the two original Köchel numbers, K. 163 referring to the finale. What happened was that Mozart turned the overture to the '*serenata drammatica*', *Il sogno di Scipione*, into a symphony of the Italian overture type by adding a final movement, the result being a homotonal symphony with all three movements in D major and the latter two in triple time. The score includes trumpets and drums.

K. 162 in C major, also with drums, dates from the spring of 1773, and while it was written in Salzburg, it may have been commissioned from Milan: hence, perhaps, its three-movement form. I cannot extract much artistic interest from it, or indeed from K. 181 in D major and K. 182 in B flat major, both composed at the same time, perhaps as part of the same commission, and both again in three movements.

The earlier of the two G minor symphonies, K. 183, written at the end of 1773 in Salzburg, is a vastly different creation. Mozart never treated his most personal key lightly. (It was not his most tragic, which was D minor – a key which he left without a symphony.) G minor to Mozart was what F minor was to Haydn and C minor to Beethoven; significantly enough, this symphony was Mozart's first in a minor key, and the first truly – one might almost say, informally – expressive one. Form and structure are, in fact, more readily open to criticism than in the case of several of his earlier works – but this criticism is praise: for the first time, darkly aware of his genius, Mozart dares. And for this very reason, he partly fails where on some previous occasions his undaring talent had all too easily, and not very significantly, succeeded. Only bad composers write chronically good music. An uninterrupted flow of immaculacy is the prerogative of mediocrity: if you don't say anything, there is not much risk of spoiling anything.

G minor and D minor, Mozart's favourite minor modes, overlap in his emotions and are, therefore, often thematically associated. The first movement of this early G minor symphony anticipates themes in the *Don Giovanni* overture and, of course, the Piano Concerto in D minor. The 'restless syncopations' (Einstein) on the tonic with which the concerto opens resemble the corresponding theme of the present work – the only two occasions, these, on which Mozart opens with a theme of this nature, mood, character and structure.

So far, when talking about backgrounds and foregrounds, we have gone into the backgrounds of the symphonies under discussion, but confined ourselves to explaining how their foregrounds came to turn into backgrounds in later, more mature works. The reason for this approach is not hard to illuminate. The foregrounds of the most immature Mozart works are so simple, so expectable, that they hardly deserve the description 'foregrounds' at all: there is so little tension between them and what lies behind them, so little does in fact lie behind them, that we might more realistically describe them as backgrounds pushed to the fore. This is the great difference between the uncommunicative music of Mozart's time and of ours; at his stage in the history of musical creation, there was a general language so that it was possible to say nothing by saying the expected, whereas in our time there is no general language, so that it is possible to say nothing by saying the unexpectable. In our own terms, the typical 'bad' music of Mozart's time is all background and no foreground, and hence insignificant; while the typical 'bad' music of our age is all foreground and no background, and hence meaningless.

With the early G minor symphony, however, this situation changes drastically: for the first time, it would be possible to define, almost throughout the work, those suppressed backgrounds against which the music is composed, thus establishing new musical meanings. An easily describable example is the beginning of the *Andante* (which, incidentally, stands in the main key's submediant).* The metrical background is indicated, in my example, by the dotted bar-lines – and, in the music itself, by the harmony, rhythm, and texture; the accents on the second beats provided by the bass and the bassoon imitations with their *fp*s leave no doubt about the two levels of the composition, background and foreground:

36

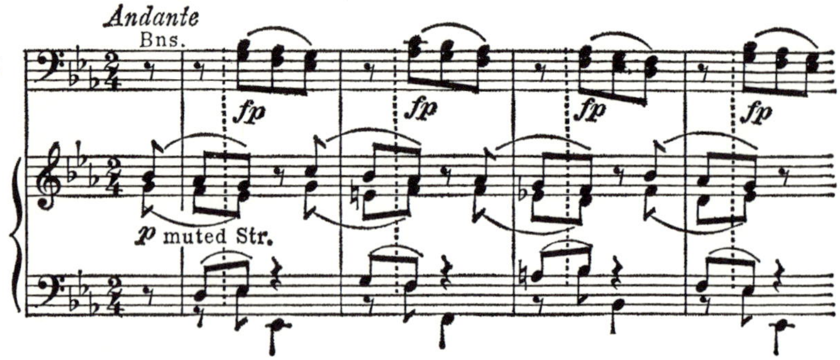

I am not suggesting that we should sit through the work with those solemn faces we normally reserve for masterpieces we do not understand; nor, however, will we get the full meaning of the music if we decide that it is 'patchy' and only switch on our attention for the striking bits: it isn't as patchy as all that. Besides, if we feel ourselves to be addres-

* The classical home for a slow movement of a work in a minor key because it is the reposeful subdominant of the relative major, which in its turn is the classical key for the first movement's second subject.

sees rather than critics, we want to know what Mozart wants to say to us not only where he succeeds, but also, if possible, where he fails. Perhaps the weakest stretches in this symphony are those innumerable semiquaver tremolos in the first movement – yet it is just there that we sense profound excitement which Mozart is as yet unable clearly to communicate. His feeling is too strong for his means of expression; he does not really know how to keep the parts moving as urgently as his feelings and indeed his own opening theme demand. In his late G minor symphony, this same kind of excitement will express itself tersely and lucidly, without recourse to pseudo-motion by tremolo. Meanwhile, we welcome the opportunity to grow into maturity with him. What is more, with our knowledge of the mature work, we shall understand the earlier communication better when we return to it: Mozart himself will have explained it to us. In a sense, it is always a great composer's more immature works that are the difficult ones. Theirs is a simplicity complicated by genius. The typical late work, on the other hand, tends to confront us with complexity clarified and indeed simplified by genius. Thus the difficulty of the typical early work is less in the ear of the player or listener than is the initial unapproachability of a late work, where the composer has done everything that can be done.

K. 184, composed in the spring of 1773 in Salzburg, is once again an Italian overture, to the extent of unfolding in three continuous movements. The slow one is in C minor: as early as his very first symphony (K. 16), we have remarked on the extraordinary relationship, in Mozart's mind, between E flat (as main key) and C minor (as slow-movement key). But the first great slow C minor movement in an E flat major work was still another four years off – the middle movement of the piano concerto K. 271.

K. 199 in G major (Salzburg, April 1773?) is a simple work, again in three movements. The outer ones are in triple time, while the middle one (*Andantino grazioso*) is not really in slow-movement mood – and so is in the dominant. It may be added that Larsen doubts the official Köchel-

Einstein date of a symphony whose first movement has, to him, more in common with the symphonies of the 1760s than with the period we are now considering. The musical interest of this chronological problem would seem to centre on Mozart's creative character itself. As distinct from the ever-advancing mind of a Haydn or Beethoven, that is to say, Mozart's creativity seems at times curiously static, so that one is unable to tell, on purely musical grounds, which is the earlier work and which the later: this is even true of some of his mature string quartets and quintets. Yet this is deceptive. It would be nonsensical to say that Mozart's was not a developing mind: what is characteristic of his type of creativity is that he appears to be able to express several stages of development at the same time, and to revert to earlier stages as if nothing, or almost nothing, had happened. To me, the most mystifying example of his timeless creative attitude has always been his last string quintet (K. 614), whose tonic movements, at any rate, I would have placed before the first of his great quintets.

K. 200 in C major was written in November 1773 in Salzburg. Its four movements, the first of which is in triple time, return us to Mozart's present developmental stage as we know it, even though the minuet is not altogether characteristic. The work is in fact more accomplished than the early G minor symphony and, proportionately, less expressive. This is not a criticism: the music manifestly fulfils its intentions. It is only when he can, in one work, reach maturity that a composer combines intentions which, at earlier stages, he has to split up between works: here is one of the reasons why, in sheer quantity, immaturity tends to be more productive than maturity.

Though only a few months separate this C major symphony from Mozart's third and last A major symphony (K. 201, beginning of 1774), the latter attains an unprecedented and, at this stage, unsuspected degree of profundity. What is more, the music is most masterly where it is most profound: talent and genius have found each other.

It is as if Mozart's 18th birthday (on 27 January of that year) had pushed him ahead of his current stage of development – by no means an absurd proposition psychologically. At the same time, Einstein is a little rash when he says that the work 'belongs to Mozart's finest creations': these, as opposed to the finest creations of most great composers, never show a weak spot.

Again, Mozart would have smiled if he had heard Saint-Foix pronounce the symphony 'one of Mozart's most characteristic instrumental masterpieces': highly characteristic it certainly is, and masterly too, but a masterpiece is masterly throughout. One has to retain some sort of meaning for these evaluative terms; otherwise, how does one distinguish between this A major symphony and any of the last four giants? Only in terms of relative maturity? That would not be fair.

Yet it is no exaggeration to call the first subject one of Mozart's greatest discoveries – maturer, too, than anything he was able to think of in any of his violin concertos in the following year. Here indeed is tension between foreground and well-implied background. So far as the rhythmic structure of this irregular nine-bar sentence is concerned, the basic motif *a*, which starts the subject off as a clear opening motif, turns into a closing and opening motif rolled into one (a^1, a^2, a^3), thus producing cumulative overlappings of phrases which violently compress the background structure – in which a^1, retaining its role as a pure opening motif, would have to enter in bar 4, after a cadential resting point in bar 3. This kind of simple complexity is characteristic of the maturer Mozart. The complexity lies in the compression; its simple expression depends on what is compressed into what: a three-bar phrase into a two-bar phrase. (Other things being equal, two-bar units are easier to apprehend than the more irregular, less dance-like three-bar units.)

Together with this compression of rhythmic structure goes a violent contradiction of harmonic background which only a pedant would bother to explain. If we cover up the bass-line and imagine a 'normal' harmonization of the tune, we get a palpable measure of the harmonic tension between foreground and background:

37

This level of inspiration, invention and construction is not sustained throughout the work, nor indeed in the movement itself. Apologetic semi-quaver string tremolos soon appear, though they are far more part of the structure than in the case of the early G minor symphony. In fact, at the end of the transition to the second subject they produce a meaningful complication, such as the very simple background of this passage would never have dreamt of without Mozart's help. By way of overlapping parts, that is to say, the second violins, apart from pursuing their own course (*a*), contribute upbeat motifs (*b*) to the principal part in the first violins:

38

If this kind of double meaning is more readily appreciable when one actually plays the work than when one merely listens to it, that only shows how close to chamber music symphonic thought still was at that stage: in chamber music, the composer addresses himself primarily to the player. A similarly expressive overlap occurs at the beginning of the minuet:

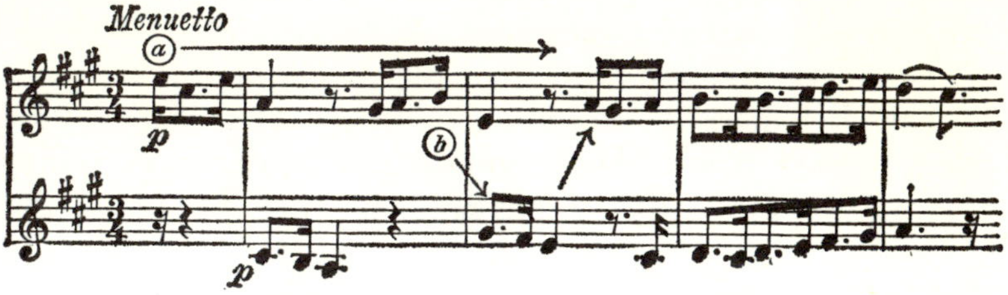

39

But on the whole, the later movements cannot compare with the first, even though there is great beauty in the fully developed slow movement, and though thematic integration has now become very conscientious – perhaps, in fact, a little too much so at times. Thus the first and second subjects of the *Andante* insist on almost the identical dominant-tonic upbeat (the second, incidentally, anticipating the corresponding theme in the first movement of the violin-viola *Concertante*, K. 364, of which more below); and the first subject of the finale, (again in developed sonata form) reverts, by way of a somewhat more impressive variation, to the octave leap of the work's opening (Ex. 37(*a*)). On the other hand, the conventional triplets of the *Andante*'s codetta come in from nowhere and only justify their existence subsequently, in the development. Mozart is here trying to introduce novel material at the end of the exposition, and so establish a close link with the elaboration – a daring stroke which, for the time being, fails, but which we yet must not denigrate; for twelve years later, in the second of the two piano concertos in A major, the selfsame approach was to result in one of the most sublime fusions of exposition and development.

As in K. 200, the minuet itself is the least characteristic movement. Most embarrassing, to my ears, are the octave unisons between oboes and horns: Mozart was no good at Haydn's (and Beethoven's) type of jocularity; he had humour instead. Or is it that I fail to hear the joke?

In the following May in Salzburg, Mozart wrote the

D major symphony, K. 202. It is the last symphony before the long pause: he did not write another symphony until the spring of 1778. Was this a real, creative, symphonic crisis? K. 202 would seem to suggest that it was; musically, it definitely 'precedes' K. 201. Almost the only characteristic of the four-movement work that places it where it chronologically belongs is the continued tendency towards thematic over-integration; a comparison between the openings of the first and last movements needs no comment:

40

It may be significant, too, that as opposed to the early G minor symphony and the just-discussed A major one, the present work does not seem to have met with the composer's approval later on in Vienna, where it remained unperformed.

The first symphony after the great pause was again in D major – as if to close the gap. It was K. 297, the famous 'Paris' symphony, written in June 1778 in Paris. For the first time, Mozart was able to conceive a symphony for large orchestra complete with clarinets: he had encountered the instrument, which he was to make so very much his own, during his preceding visit to Mannheim – a visit he had extended, against his father's wish, because he had fallen in love with Aloysia Weber. The Mannheim influence is indeed strong in this music. True, Larsen writes that 'this is the first really mature symphony in the Viennese classical style'; but most of us will find it easy to concur with H. C.

Robbins Landon, who suggests that it 'is not really a Viennese classical symphony at all, but rather a conscious attempt to write an orchestral work in the grand Mannheim style: it is, therefore, more an imitation than a true assimilation, and in K. 318 [in G major] and K. 338 [in C major] the "grand manner" is far more personally (and thus more convincingly) set forth'. There is a cold brilliance about the masterly aspects of the work, and the delight with which musicologists concentrate on its undoubted historical importance tends to show that it has not altogether come from, or gone to, the heart.

At the same time, there is some musical significance, so far neglected, in one particular historical event in the first movement – the absence of the customary repeat. It is true that formally, this apparent innovation is achieved by a reversion to older models rather than through the evolution of sonata form which came to produce the revolutionary repeatlessness of the first movement of Beethoven's string quartet, Op. 59, No. 1 in F major. Nevertheless, the 'Paris' movement conforms sufficiently closely to the sonata scheme to make the omission proportionately unexpected and striking, and so to justify the assumption that Beethoven himself may have received his structural stimulation from Mozart's example. This, no doubt, was never intended to be formal news, except perhaps somewhere at the back of Mozart's creative mind. How strongly the concept of the repeated exposition is ingrained in first-movement form can be judged from the fact that Beethoven's recipe for 'overcoming' the omission of the repeat continued to be followed by composers at later stages in the history of sonata form. What he did was to return to the first subject in the tonic at the beginning of the development, and thus to build a *mental bridge passage* between the expected repeat and the development actually ensuing. We encounter this device as late as Schoenberg's last tonal sonata work, the second string quartet in F sharp minor, whose first movement, though in radically modified and compressed sonata form, is still composed against the background of an expected repeat: the foreground tells you where the repeat, the double-bar, would have been. Now, Mozart's repeatless movement also reverts to the first subject at the beginning of the middle section, though the theme appears in the dominant in a way reminiscent of older forms. Nonetheless, one can imagine Beethoven composing his tonic 'bridge passage' against the background not only of his own omitted repeat, but also of Mozart's corresponding resumption of the first subject.

Mozart wrote two slow movements for the 'Paris' symphony, and until fairly recently it was thought that one was the other – with delightful results in our musicological and critical literature, where the interested reader will find words too tough to be subsequently eaten. The facts, simply, are these:

The original slow movement is the one we all know:

41

It did not please its customer, the conductor Le Gros, at the first performance of the symphony in June 1778: he thought it too long and too modulatory. Mozart, though he regarded this criticism as rubbish, obliged with another slow movement for the second performance in August:

42

This replacement was published in parts by the Parisian firm of Sieber in 1789; it is only included in one modern edition. Now, Eric Blom says Mozart said 'he liked both [movements], each for its own peculiar qualities'. Mozart's actual words, however, give a different impression (letter to his father of 9 July 1778, in my translation):

> 'Each is right in its way, for they have different characters; however, I like the last [Ex. 42] still better....'

In other words, although Mozart had not accepted the need for improvement, he did, in the event, prefer the result. The obvious conclusion has never been stated – probably because people still are not quite sure which movement is which: *we never hear the 'Paris' Symphony with the slow movement which the composer preferred.* I would suggest regular performances of the 'Paris' symphony in the version which Mozart preferred; the work gets enough performances for both versions to become, and remain, equally familiar.

The last movement, *Allegro*, is certainly the most original. Where Mozart offers historical precedent in the first movement, he provides his own structural fulfilment in the finale. My 'where' is not rhetorical, but literal: I am again concerned with the beginning of the development section which, to my mind, is the one profound master-stroke in the work – the accent being on 'profound'. Einstein seems to miss the more important half of the truth when he writes: 'The second theme of this movement is a *fugato*, supplying the natural material for development; it does not return in the recapitulation – one of the strokes of genius in this masterful movement...' The stroke of genius lies elsewhere; Einstein's observation merely concerns one of its consequences. What Mozart does in the first place is to repeat this first theme of the second subject in its original (dominant) key at the beginning of the development, thus necessitating its omission from the recapitulation. In terms of the resultant tension between his foreground and his implied background, he has thus created a complexly meaningful situation. For one thing, far from reverting to older forms, as in the first movement, or retaining newly established forms, he composes straight against them all: *instead of the first subject in the dominant or the repeat of the first subject in the tonic*, he introduces a *repeat of the main theme of the second subject* in its rightful *dominant*. It will be noted that the closest background to this event is his own first movement: there, in the parallel place, we get a dominant restatement of the first subject. In the finale, this is replaced by a dominant repeat of the second subject's main part. (We should never, I think, speak of a 'repeat' or 'repetition' where a theme is restated in another key.)

For another, and perhaps even more important thing, the repeat of the second-subject theme at this point cannot but create the impression of a recapitulation in what would thus be an A–B–B–A form – an impression reinforced, despite the ensuing development of the theme, by its subsequent omission from the recapitulation proper. Mozart so achieves and produces the experience, well-defined to any observant listener, of an interpenetration of formal sections: sonata form (including that of his first movement) is his background, a rearrangement of the sonata elements his highly inventive foreground. We shall see later how, as a mature symphonist, he subtilizes this process as a result of richer emotion, thus heightening the tension of his music.

For the moment, it merely remains to be added that after the first performance of the 'Paris' symphony, Mozart was so overjoyed with the applause that he immediately went to the Palais Royal, ate a first-class ice cream, prayed as he had vowed to do (in that order, apparently), and returned to his quarters (letter to his father of 3 July).

K. 297b, the *Sinfonia concertante* originally scored for flute, oboe, horn, and bassoon, with orchestral accompaniment, was also written in Paris, between 5 and 20 April of the same year. It really falls outside our terms of reference – unless we include all the concertos, not to speak of the string quartets, quintets, etc., in our survey of Mozart's symphonism. Nevertheless, to those who expect to read something about the work in this chapter, one or two points of factual interest will perhaps be welcome.

First of all, the original score is not extant (though one still hopes that it may turn up); all we have is an arrangement for oboe, clarinet, horn, and bassoon, whose authenticity is more than doubtful. Secondly, there is a more purely musical fact of considerable fascination which, so far as I am aware, has never received attention: rather exceptionally for the grown-up Mozart, the three-movement work is cast in a homotonal frame – E flat major. Now, this was the time when he wrote his so-called 'Mannheim' sonatas for piano and violin. There are seven of them and five are, in formal layout, modelled on Christian Bach, who was very much in Mozart's mind in Mannheim and Paris. Thus they are in two movements and hence inevitably homotonal. The piano sonata with the Turkish march (K. 331), likewise written at the stage, is homotonal too (A major and minor), although it is, of course, in three movements. In short, one way or another, homotonality was momentarily in the air for Mozart at this particular point in his development; and it is astounding how he achieved intense variety within the tonally unvarying framework – so much so that the framework, in the case of each of these works, has gone unnoticed. Several years later, in 1781, he was to return once more to the two-movement scheme with its automatic homotonality – in the Sonata for piano and violin in G major–minor–major, K. 379. But the three-movement homotonality of the wind *Concertante* or the 'Turkish' sonata were to remain his most extended achievement of the kind; unlike the mature Haydn, the mature Mozart never came to write four movements without changing the keynote.*

Although K. 311a, an Overture, is a highly doubtful work in the version we know, Einstein suggests that this may be a later arrangement of an authentic Paris work of the same year (August or September).

K. 318 (Salzburg, 1779) is another, far more famous overture – the G major one 'in the Italian style', as it is generally called. Einstein thinks that Mozart wrote it for the opera *Zaide*.

K. 319 in B flat (Salzburg, July 1779) is a most impressive work, even though there are triplets all over the place to keep up the movement: as a symphonist, Mozart had not yet

* See also p. 58.

arrived at the stage where he could always rely on his ideas propelling themselves through characteristic rhythms and their developments.

Originally, the work had had three movements; the minuet appears to have been composed in Vienna three years later – which is probably why it 'fits' rather self-consciously, its theme (Ex. 43) being a little over-integrated with the openings of the outer movements (Ex. 44 and 45 respectively):

43 

44 

45 

Only once, much later, did Mozart sublimely succeed in establishing close natural relations between the themes of different movements – in the D minor quartet, K. 421 (1783), perhaps significantly one of the set he dedicated to Haydn: it was Haydn, rather than Mozart, to whom extreme economy of the basic thematic material came instinctively.* For the rest, it is the homogeneous mood of the D minor quartet's three tonic movements that naturally makes for thematic homogeneity. Now, while I am the last to incline towards seeing chronological connexions between life and

* See p. 60.

art, the curiously persistent mood of dramatic and indeed optimistic anxiety that informs the quartet may, I think, have something to do with the probable fact that Mozart wrote this music during his wife's first delivery.

K. 319's first movement is repeatless, like the 'Paris' Symphony's. Or perhaps we should say – unlike the 'Paris' Symphony's. For hardly any of the structural features recur here which made that earlier absence of the exposition's repeat so interesting; nor are they tangibly remembered and replaced. One of the most forceful reasons for repeating the exposition lies in the development of its material – some of it, anyway – in the middle section: that which has to be developed has a natural need to 'sink in' in the first place, and repetition is the most immediate means towards that purpose. In K. 319's opening movement, however, the development section concentrates on material which has not been expounded at all, and the need for the exposition's repeat is weakened accordingly; in fact the exposition itself is so constructed as virtually to stifle any expectation of a repeat in the first place. On the other hand, as soon as we hear the theme of the development, we realize that it does not stand in need of exposition:

46

It is in fact a tag, first used by Mozart, in the same key, in the *Credo* of the Mass in F major, K. 192 (1774) – and last, of course, as basic motif and fugal subject of his last symphonic movement:

47

Now, it may be said that in view of my argument, there is not much point left in the repeated exposition – i.e. orchestral and solo exposition – in a concerto movement like the opening one of K. 488, the 'big' A major concerto of 1786, where the main material of the development is also a new theme which forms the link between exposition and development:

48

The orchestral exposition, however, is the basis, not, in the first place, for the development, but for the solo exposition, which is a repetition only of outlines: within them, there is intense variation, both thematic and harmonic.

Again, it might be objected that the finale of our present symphony gives the lie to my argument: although the development centres once more on a new theme (Ex. 49), the exposition is, this time, repeated.

49

The answer is threefold. In the first place, this is one of those movements where both parts are repeated, because without this balancing device the music, which moves at *allegro assai* pace, would simply be over too quickly. In the second place, the development's theme is in fact introduced in view of this 'repeat' structure: unlike the corresponding theme in the first movement, it is not a tag and adds new thematic interest where otherwise we might feel we are hearing too much of too little a thing, for the material of the movement is, after all, slight. In the third place, Mozart heightens the function of the exposition's repeat by yet including more of its material in the development than he would otherwise have done – more, in fact, than he does in the first movement.

K. 338, composed in Salzburg in August, 1780, is Mozart's last C major symphony before the final 'Jupiter'. As it stands, it is another three-movement work. But for one thing, in the autograph, there is the beginning of a minuet, subsequently deleted. For another, Einstein points to the likelihood of the hardly-known Minuet, K. 409 (Ex. 50), having been written for this symphony at a later stage (May 1782).

50

I would support this view, on purely musical grounds, because there is tangible background homogeneity between this movement and the rest of the work. I join Einstein, then, in pleading for the inclusion of the movement, which I would play before the *Andante di molto*, both because Mozart planned his original minuet in that place and because the effect of the slow movement being followed by the finale (*Allegro vivace*) ought perhaps to be retained. The first movement, another *Allegro vivace*, is again repeatless, but *mutatis mutandis* it is as unlike the corresponding movement of the 'Paris' symphony as the first movement of K. 319: the development's material is not stated in the exposition. Perhaps however, before we take leave of this kind of sonata structure, we ought to remind ourselves that the 'new' material in such contrasting development sections is not, of course, new at the background level; if it were, the movement simply would not hang together – and that goes for simple ternary movements too. The surprising, almost shocking beginning of the present development section (Ex. 51), for example, evinces a readily audible background relation to what has just been heard in the exposition (Ex. 52); by way of change of key and mode, the one represents a retrograde version, harmonically anchored, of the other:

51

52

The A major fragment K. 320e, called *Sinfonia concertante a tre Stromenti Violino, Viola, Violoncello*, and reputedly composed in the summer or autumn of 1779 in Salzburg, was completed by Otto Bach (1870–1). My attention was first drawn to this reconstruction by the viola player Herbert Downes, and I mention it here because it does not yet seem to have been performed in this country; there is enough real Mozart here to make an occasional performance desirable.

The more famous *Sinfonia concertante*, that for violin and viola (K. 364), really bears the German title *Konzertante Sinfonie*, which has to be replaced by its Italian equivalent, as there is no English word for *Konzertante*. Probably written in the summer or late summer of 1779 in Salzburg, this E flat major work – with an *Andante* in C minor* – is a major masterpiece, greater and deeper than any of the fiddle concertos: Mozart had not only grown maturer, but was, I think, more inspired by the viola, which he liked playing, than by the violin, which he didn't. Like the wind *Concertante*, K. 297b, the piece should not really be included

* Compare our remarks on K. 16 (p. 54) and K. 184 (p. 65).

in our present discussion: it is a concerto. But it does contain a problem to whose solution we may perhaps sacrifice a little space.

I mean the question of the so-called *scordatura*, i.e. the tuning of the viola a semitone up, in order that the part be read and, so far as fingerings are concerned, played, in D major. Almost invariably, the present-day reaction is to disregard Mozart's instructions, leave the tuning of the instrument intact and play in E flat major. The underlying assumption, correct so far as it goes, is that viola technique has sufficiently improved to make a good performance in the difficult key of E flat possible. However, the most important aspects of the *scordatura*, which this approach ignores, is that the colour of the open-string key D major is different from that of E flat, and *that it retains this difference when the strings are tuned up a semitone so that, as it were, D major is played in E flat*: the colour depends not on the absolute pitch of the key, but on the silent, over-tonal contribution of the open strings as well as the technical framework for fingerings. In other words, Mozart wrote the 'Concertante' bi-tonally in one key so far as colour was concerned: to the more brilliant violin he gave the subdued key of E flat, while to the more subdued viola he gave the brilliant key of D, thus establishing a balanced texture between his solo instruments. Now, while it is true that the modern viola player can achieve more brilliance in E flat than Mozart might have thought possible, especially when he plays on one of those outsized modern instruments that don't sound like violas anyway, there is, I submit, no question that on balance the truest sound-picture is obtained if Mozart's *scordatura* remains heeded. Nor, incidentally, should we forget that the sudden sharpening of an instrument's strings produces a more penetrating sound anyway, which Mozart must have had in mind too. (For once, then, I find myself in the camp of the historical faithfuls, except that they have never yet discussed this particular problem.)

With the 'Haffner' symphony in D major (K. 385), of

1782, so called because it was written in Vienna for the Haffner family in Salzburg, we reach the first of the last: only the 'Linz', the 'Prague', and the final trilogy of 1788 were to follow. We realize, then, perhaps a little more forcefully than when we opened our investigation, how relatively inessential was purely orchestral symphonism to the development of the most essential Mozart: when all is said and done, he only wrote, at the most, six great symphonies in his life, i.e. the last six, which are 'great' in the sense in which we speak of a great life-embracing symphony in the case of every other symphonic master from Haydn onwards. As we indicated at the outset, the last four are the greatest.

When people speak of jaded musical palates, they usually think of the contemporary lack of appreciation of the finer harmonic points (whether expressed vertically or horizontally) in the music of the past. I have never yet encountered a complaint about rhythmic insensitivity, although it is with the subtle complexities of rhythmic structure (harmonic rhythm included, of course) that classical music stands or falls. Anybody, for example, who is not violently struck by the regularized irregularity of the antecedent of the 'Haffner' symphony's opening theme (Ex. 53(*a*)), has no chance of following the full meaning of the further course of the movement.

53

Superficially, the five-bar phrase extends the more regular, sequential four-bar background No. 2 (Ex. 53(*c*)). But (as I hope to have demonstrated elsewhere) every good extension hides, and explains, an underlying compression; for to extend for the sake of extending is unfunctional, unfactual, insubstantial, in any form of communication – an empty hypnotic device which at early cultural stages has some ritualistic significance, but which in our civilization only survives in political speeches – the most primitive form of modern human communication, advertising apart. In short, the ultimate background level of which Ex. 53(*c*) itself is a compression is the symmetrical eight-bar scheme of Ex. 53(*b*). Without going into the details of how I reconstruct these backgrounds, which in any case only serve their purpose if they instinctively 'click' with the reader, attention might be drawn to the fact that where different background levels exist, each of them will always be found to be directly represented in the foreground. Thus, while background No. 2 in Ex. 53 is more directly related to the foreground than is background No. 1, one element of background No. 1, i.e. the rhythm of the opening octave, is used in the foreground towards the extension of background No. 2. And the latter's opening rhythm, in its turn, makes its foreground appearance at a later stage:

54

For the rest, the proof of the antecedent lies in the consequent – which is what is meant by balance: the eight-bar scheme of the opening theme's consequent (Ex. 55) answers the eight-bar scheme of the antecedent's original background (No. 1 in Ex. 53).

55

If question and answer proceeded altogether on the same level, if eight-bar phrase simply answered eight-bar phrase, there would be a corresponding loss of tension and indeed of musical meaning.

When discussing the finale of K. 297 (the 'Paris' symphony), we had occasion to point to a relatively immature example of Mozart's interpenetration of formal sections, with sonata form as his background, and a rearrangement of the sonata elements producing his foreground, his structure. In the first movement of the 'Haffner', this process reached a degree of subtlety and perfection that was to impress itself on the minds of even the distant future's great creators – as we shall see in the case of the Fourth symphony of Tchaikovsky (p. 347 f.), who was a deep and fruitful admirer of Mozart. The juncture I am thinking of in particular is Mozart's second subject. I say 'juncture' rather than 'place', because the point is that the second subject does not have any single place in the exposition. Thematically, the one decisive contrast is Ex. 56, and even there the first subject, which indeed comes to dominate the entire movement,

56

continues underneath in the violas – a circumstance which has led many to describe the movement as 'monothematic', as if to show that they can hear more than one tune at a time. Now, it is this new theme which virtually all commentators call the 'second subject'. To ears harmonically alive, however, the expected characteristic of the second subject is not just thematic contrast but, more basically, the well-established dominant. And from this harmonic point of view, Ex. 21 is part of the bridge passage, leading as it does from the dominant's dominant (the dominant's establisher-in-chief) to the dominant, upon whose establishment Ex. 54 ensues. Yet this passage, far from yielding a new theme, already invests the first-subject material with a marked *codetta* feeling, which the ensuing minor-mode variation does nothing to weaken. In other words, when the structure arrives at Ex. 54, it says 'I've missed it' (i.e. the new tune in the dominant), and adjusts itself accordingly. Or more seriously speaking, the two functions of the second subject – harmonic and thematic – are split, with the thematic second subject penetrating the harmonic stage of the transition, and the harmonic second-subject stage serving as basis for the first steps of the gradual thematic rounding-off, or rounding-up, of the exposition. This is live structure as opposed to – as composed against – dead-alive form.

The remaining movements are simpler though no less inspired, but they do remind one that the symphony was composed as a serenade in the first place, complete with an introductory march (K. 408, No. 2) whose theme (Ex. 57) betrays its original context, and another minuet,

57

which seems to have been lost. Nor was this the only time that an intended light-weight work grew into one of symphonic proportions: the late Divertimento for string trio, K. 563, met with a similarly fortunate fate in the process of creation – more fortunate in a way, for there Mozart succeeded in retaining all his six movements without harm to the symphonic stature of the work.

The 'Linz' symphony in C major, K. 425 (Linz, November 1783) is the 'Paris' symphony turned warm. Like the latter, it was composed as an occasional piece, 'head over heels', as Mozart put it; but now the brilliance was matched by depth. It has often been said that the work shows no trace of urgency, what with the profound slow introduction (a rarity in Mozart, surpassed only by the introduction to the *Don Giovanni* overture, which unfolds similar chromaticism) and the general perfection of it all. I would, on the contrary, suggest that once again life and art are immediately connected: the music evinces every trace of urgency, but on a plane so highly sublimated that no sooner does one understand that it has happened than one fails to understand how it can have done. The main themes of the outer movements, (Exs. 58 and 59(*a*)) in particular, are

58

59

amongst the most elatingly urgent melodies Mozart ever wrote – as is the finale theme of the 'Prague' (Ex. 60), the next symphony Mozart was to write, though another three years elapsed before he did so. In either finale, moreover, I detect a Hungarian gipsy tinge (marked '!' in Exs. 59(*a*),

60

(*b*), and 60) most unusual in Mozart's style, as opposed to Haydn's; this 'foreign' element is quite clearly used towards the expression of unexpected urgency. The *poco adagio* is composed against a *siciliano* background – a forerunner of the slow movement of the piano concerto in A major, K. 488.

K. 444 in G major is not by Mozart; only the twenty-bar introduction is: he wrote it for a symphony by Michael Haydn – at the same time, and for the same Linz concert, as the 'Linz' symphony.

And so to the 'Prague' symphony (K. 504) itself, one of

Mozart's most towering achievements; he composed it towards the end of 1786, and first performed it in Prague in January, 1787.

Earlier on, we noted that the three-movement scheme is, in itself, more primitive than the four-movement one, which emerged under the Viennese influence. It is, I think, a natural law in art that primitive forms are reverted to long after they have been abandoned, and enriched by the creative experience meanwhile accumulated. It is as if the artist felt under an obligation to return to old backgrounds as soon as they are capable of yielding new meaning: in this sense, every communicative artist is an innate conservative. Schoenberg returned to tonal backgrounds, *inter alia* in the *Ode to Napoleon* (E flat), although he hardly noticed that he was doing so and certainly didn't know why. We shall soon see the same law operate, in yet two other dimensions, in the finale of Mozart's last E flat symphony. In the case of the 'Prague', *the law turns what was a scheme without sonata movements into one without anything else*. This, I suggest, is the simple explanation of why there is no minuet. All three movements of the 'Prague' symphony are in fully worked-out sonata form; it is, in fact, the only classical symphony which, dispensing with the most striking formal contrast of symphonic tradition, celebrates an exclusive sonata festival in which new and highly original contrasts of mood and rhythmic structure take the place of the established contrast between the minuet and the rest of the symphonic build-up – on the basis of the pre-established three-movement scheme. Sonata form had to reach a high degree of evolution, and of proportionate differentiation, before this master-stroke of formal economy could become possible. Two minds, and two minds only, were responsible for this vast and fast flowering of what, potentially, has remained the most meaningful instrumental form – Haydn's and Mozart's. There are minor contributions from other sources, so beloved by musicologists, but on inspection they prove dispensable.

As a musician-writer, one approaches the final trilogy –

K. 543, 550, and 551, created within seven weeks and three days in the summer of 1788 – with a certain amount of well-considered trepidation. The 'Prague' symphony may still conceivably be in need of stimulating comment, but the last three are as popular as they are great, while at the same time volumes have been written about them at all levels of expertise and impressionistic criticism.

When I commented on the early G major symphony, K. 110, I pointed to the fact that the opening 3/4 *Allegro* forms a remote background to the first movement of the E flat symphony, K. 543. That use of the term 'background' might seem different from its use in later contexts, where 'backgrounds' either appear in the same works as their foregrounds, or do not appear at all, but are only defined by foreground implication. In creative reality, however, all these backgrounds have the same significance: they are always audible or divinable when their foregrounds are being heard. In formal essence, that is to say, that early background of K. 110 exists in your mind when you hear the E flat symphony, whether you actually happen to know K. 110 or not; and likewise, it would have existed in Mozart's mind even if he had not written the early work. Without such implicit terms of reference, music would remain incomprehensible, a private language.

As for K. 543's triple time itself, exceptional in first-movement form at this late stage in Mozart's development, our law of the enriching return to the primitive again seems to be in operation. We have noted previously that triple time offers the immature creator an overdraft in terms of characterization of motifs and phrases. When the mature, solvent creator returns to the scene of the transaction, he is able to turn the overdraft into a credit account. The self-propelling, characterizing power of the triple metre is now scorned and contradicted, turned once again into a background, against which a highly original rhythm develops (Ex. 61). Though starting on the main beat with the tonic, it takes the shape of a twice-postponed upbeat until the first proper downbeat falls on the fourth bar of the four-bar phrase:

61 

The effect of this thematic character is, of course, heightened by the preceding *Adagio* introduction in *alla breve* time.

The interpenetration of sonata elements which we last discussed in the first movement of the 'Haffner' reaches a yet higher degree of refinement in this first movement, and at the corresponding juncture. Gordon Jacob says that the second subject 'appears' at Ex. 62. As it stands, the statement is wrong. The second subject, once again split in its functions, emerges in stages, of which Ex. 62 is certainly not the main one. True, it is a new melody, and it does not modulate, as

62 

it does in the 'Haffner': the dominant's dominant this time clearly precedes it. Nevertheless, through its thematic character and its harmonic position, Mozart achieves a marked overlap with the bridge passage to which it comes to form an afterthought: its cadential nature and the fact that the dominant has only just been reached, by no means over-decisively so, turn the passage into a *codetta* of the bridge passage (to coin a term), which confirms the arrival in the dominant. The main stage of the second subject is thus reached with Ex. 63 which, as distinct from Ex. 62, is a proper

63 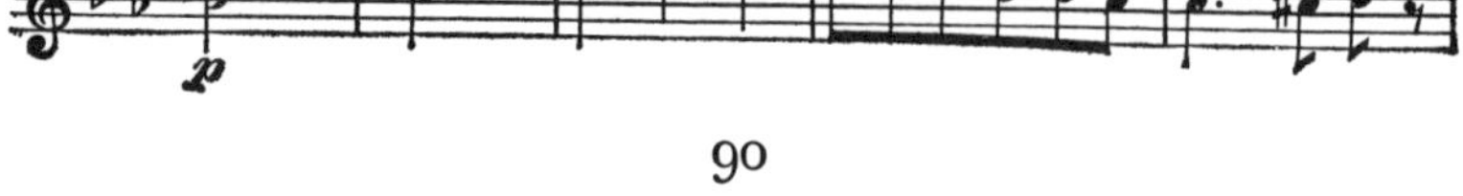

tune deeply embedded in the dominant. Yet, as in the 'Haffner', the structure has almost missed the second-subject boat, and its forward-urging intensity is proportionately increased: by now, real *codetta* feeling is accumulating, and there is no time to spread oneself, second-subject-like, in the dominant. In fact, all there is room for is a 5 + 5-bar(!) period (beginning at Ex. 63), with whose last bar the first stage of the *codetta* proper overlaps:

64

Gordon Jacob's analysis appears in the widely read Penguin Score of the work, and its more preposterous findings must therefore not remain unnoticed. He says that the second subject consists of two contrasted ideas, Exs. 62 and 64. Of Ex. 63 he does not breathe a word, so that we are led to assume that Exs. 62 and 63 are a single idea! Nor does he seem to notice that Ex. 64 resumes the first subject's material which has meanwhile been developed in the transition: it isn't a new idea at all, but is on the contrary one of those bridging, get-a-move-on patterns predestined to work up *codetta* feeling. But then, for him, 'the exposition ends with a short *codetta* of eight bars'.

Nevertheless, the wrong observations of professionally trained analysts are always interesting, for the composer will usually be found to have had a hand in them: they tend to occur when an interpenetration of formal elements has taken place, whence the form is mistaken for the structure, the background for the foreground that is composed against it. Thus, the analysts describe meaningfully what the music is *not*; only, they don't say so.

The *Andante* theme is a characteristic example of Mozartian irregularity (alternating $1\frac{1}{2}$-bar and $2\frac{1}{2}$-bar phrases) wrapped up in a symmetrical parcel (4 + 4 bars) for easier acceptance, whereas the minuet is perhaps the only great Mozart movement with a low degree of hidden complexity, palpably designed as a relieving contrast – as very much distinct from, say, the minuet of the G minor symphony.

In the *Allegro* finale, we again encounter the law of the enriching return to the primitive in a new dimension – that of thematic organization. Already in the opening movement of the 'Haffner', a certain urge towards monothematicism had made itself felt, and while we could not say that the movement is based on one theme alone, the fact remains that the basic theme is never absent from it. The present finale, on the other hand, is virtually monothematic. Sonata form, which had started monothematically but was responsible for the development of polythematicism, here returns to its monothematic base – and with a composer, too, whose creative character had proved passionately polythematic, while Haydn continued to do wonders with monothematic sonata form: as we have pointed out, he preferred to be polytonal, far more so, in his turn, than Mozart.* Haydn's influence is certainly detectable in this finale, though there are mature monothematic sonata forms of Mozart's where it isn't: *The Magic Flute* overture (K. 620, 1791), for example, or that fantastic movement which has it both ways, mono- and bi-thematically, i.e. the first movement of the Clarinet Trio (K. 498, 1786), where monothematicism develops in the middle-ground, as it were, the concluding cadence of the first subject (itself based on the basic motif) being virtually identical with the opening phrase of the second. One notes that the return to the primitive develops gradually: the earlier structures are not wholly monothematic, the latest (*The Magic Flute* overture) is absolutely so. The more mature you are, the more primitive you can afford to be.

If the three last symphonies are the best-known, the G minor is perhaps more best-known than the other two. Its emotional impact is direct, its language Mozart's most personal, the key his most characteristic – only once before (in

* It will be noted that I use the term 'polytonal' in a new sense; in fact, I don't accept the old one. Two or more keys at the same time don't exist in aural reality: one is always strongest at any given juncture, and absorbs the others. 'Polytonality' can therefore refer only to successive keys.

the string quintet of the preceding year) and never again used to express symphonism of equal consequence. In the circumstances, it does not surprise us that the symphony is also one of Mozart's most advanced compositions, more 'modern' in our sense than any of his other symphonies – harmonically in the first movement, as Schoenberg has shown in his *Harmonielehre*; melodically, too, in the last, as I hope to show below.

But first of all, a word about the actual theme of the first movement – one of the most original that Mozart conceived, most tense emotionally, and hence most stringent in organization.

Haydn was the first to start a sonata form before the theme, i.e. with the accompaniment (cf. the so-called 'Bird' quartet of 1781), thus increasing the span of the sonata arch. But it was left to Mozart to introduce textural movement into the device (without, of course, changing the actual harmony at this preparatory stage), an approach adopted, amongst others, by Schubert in his A minor quartet and Mendelssohn in his violin concerto. One observer has gone so far as to suggest that the first three beats of the work leave the key open: since the D does not enter until the fourth beat, the first three could conceivably be in E flat:

65

It needs an astonishing degree of harmonic deafness or at least perversion to arrive at this impression. A good composer expects every harmony to be instinctively interpreted in the simplest, i.e. strongest possible way: if a third, major or minor, is heard, the lower note or bass is intended to be understood as the root of a triad (if it is not contradicted, and even where it is not confirmed in the bass, as it is in Ex. 65), just as is the upper note of a sixth. Nobody has yet suggested that the afore-mentioned 'Bird' quartet starts in A minor:

66 

No, in either case, the fifth enters merely to complete the triad and thus herald the full beginning of the piece.

Of Mozart's melody itself, Erwin Stein writes in his posthumous *Form and Performance* that it 'centres on the leap of a sixth. We recognize the line's direction upwards in disjunct motion and, subsequently, downwards in conjunct motion.' This is perfect description which, however, needs analysis. The sixth is central not only as peak and middle of the phrase, but also because after it has happened melodically, the conjunct motion, which had dominated the antecedent too, can never be the same again: the thirds implied in the consequent's downward flow are composed, responsively, against the background of the sixth, of which they are inversions, and which, of course, appear in the introductory accompaniment in the first place:

67 

This is what I mean by stringent organization under high emotional pressure: basically, the theme consists of only two intervals.

As for the theme of the slow movement (Ex. 68), it has been said that it hides that famous tag which Mozart used as his basis for the finale of the 'Jupiter' symphony (cf. Exs. 46 and 47). The observation may seem artificial; it seemed so to me, at any rate – until I changed my mind. After all, the numbered notes in Ex. 68 are the main ones

68

from the harmonic point of view, and this is a theme which, up to the end of the antecedent, is produced by the harmony. If the German theorist Heinrich Schenker (about the only analyst whom Furtwängler ever accepted) had reduced it to one of his 'basic lines' (*Urlinien*), our tag would have emerged even if he had never heard it before. Was the 'Jupiter' already simmering? After all, these three symphonies were written in one breath and show, from work to work, a contrasting continuity which we would expect in Beethoven, but do not find in Mozart elsewhere – not even in those six string quartets which he dedicated, as one work as it were, to Haydn. As we shall see, the tag makes another subterranean appearance in the 'Jupiter' itself before it emerges as Mozart's final symphonic theme: the possibility of a gradual and probably unconscious – but none the less purposive – journey towards it cannot be denied.

The minuet (*Allegretto*) is written against the background of a minuet – more violently so than any other of Mozart's. Eric Blom, to be sure, calls it 'smiling rather than rollicking music'. I cannot hear much of that smile, except in the trio and perhaps in the last six bars of the principal section, which in any case are by no means untinged with sadness. I may, of course, be misunderstanding Blom in the sense in which I say we are misunderstanding, or not understanding, Schumann's impression of the whole symphony; on the other hand, 'smiling music' does not readily proceed by way of a continuous *forte*, masculine and intermittently polyphonic.

In the finale (*Allegro assai*), whose first eight-bar period Mozart wrote out twice for the sake of stressing the balance of the theme, where otherwise a repeat sign would have done, there is one shattering moment of threatening disintegration: the beginning of the development, where the theme is torn to meaningful shreds, only to be hastily reassembled. The emotional field here explored had not previously been approached by any composer, and it behoves us to reflect 'how Mozart does it'.

Luigi Dallapiccola has described how he was impressed by Heinrich Jalowetz's observation about the quasi-dodecaphonic nature of this passage, where ten different notes appear in pretty 'atonal' succession. Misleadingly, Dallapiccola speaks of a 'series' of ten different notes: it is not this panchromatic arrangement of the notes that is used serially – although a much smaller group of notes is. Our twelve-note experts, even those in the highest circles, have been too busy counting notes in classical music, instead of listening for the very technique whose origins they have been trying to discover. Admittedly, the appearance of most or all different notes in a classical line is important enough, but its importance is purely negative: it signifies an anti-tonal disintegration of a structure whose context is tonally organized. The yet more important, positive question immediately arises: how is the music re-integrated? How is the tonal and indeed rhythmic disintegration counteracted? By a tiny motif which assumes the function of a tone-row, consisting of the diminished fourth followed by the diminished seventh (BS, i.e. 'basic set' in Ex. 69).

From now on till the end of this paragraph, the interested reader is invited to follow my demonstration slowly – while the uninterested reader will most comfortably survive if he turns straight to the next paragraph. Well then, it must be borne in mind that the serial re-integration has to cope with a double task, namely, to secure the unity and continuity of the anti-tonal passage itself, and to secure the unity and continuity between this passage and its tonal context. Accordingly, we find that the basic set is already latently present before the tonality disintegrates, i.e. in the notes I have marked BS° in Ex. 69; it is really here that the strict serial organization starts:

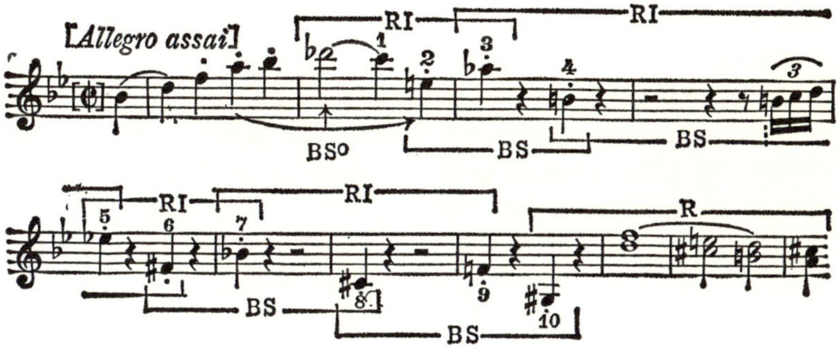

69

The blow to the tonal organization begins with the last note of BS° and the ensuing diminished fourth which is already the first interval of the basic set's first manifest entry (BS), as well as being the second interval of its first, as yet half-latent retrograde inversion (RI): serialism has the situation well in hand before the tonality and rhythmic structure are attacked. And at the far end of the passage, both unity and continuity with the resumed tonal organization are again secured by an overlapping and firm interlocking of the two techniques, serialism and harmony: the 'lead-back' into the re-establishment of the key-system proceeds by way of the fifth rotation of the basic set and, overlapping with it, the first and only entry of its retrograde version (R), again half latent: manifest tonality is taking over. Of course, there is latent tonality in the whole anti-tonal passage, which can easily be harmonized – at the expense of the feeling it expresses. (Schoenberg, incidentally, once harmonized the theme of his serial *Orchestral Variations*.) It will, finally, be noted that in the harmonic sphere at either end of the anti-tonal eruption, the row is filled in by the respective tonics (B flat and D), whereas in the course of Dallapiccola's ten notes, it remains ruthlessly naked; the repeated B in the fourth bar follows, if the meaningful anachronism be allowed, Schoenberg's practice and theory.

Mozart wrote the G minor symphony in two versions – without and with clarinets.

Before we reassess the historical position of the 'Jupiter' symphony, we might devote a thought or two to the question of important historical figures, of whom Mozart is not generally considered to be one, in spite or because of his unique genius. In fact, for some considerable time now, our musical world has accepted the fact that important historical figures are one thing, great composers another. Of course, minor composers have had creative urges and ideas of major historical significance. But the probability remains that Haydn, without his strenuously famed forerunners, would have done his job single-handed; whereas without Haydn, the whole course of our musical history would certainly have been different. As for Mozart himself, he is the official arch-conservative amongst the greatest composers, yet it was to him that Schoenberg turned as a self-confessed, devoted pupil, not, with respect, to Carl Philip Emanuel Bach. We have just had an inkling of how Schoenberg may indeed have been more receptive to Mozart's procedures than he himself knew; elsewhere, I have given further examples of Mozart's serial technique which in my opinion had a definable influence on Schoenberg.

Now, at the outset of this chapter, I have drawn attention to the somewhat facile view that the 'Jupiter' symphony is an obvious end; I suggested that the 'Eroica' would have been considered just as much of an end if Beethoven had finished symphonic writing at that stage. There is another side to this argument. The 'Eroica' has always been regarded as the mother of the symphony as we know it. Most of us agree with Mahler that the symphony should be a world; and we think that Beethoven was the first to make it one. One of his basic innovations towards this end – we have read about it in countless learned pieces – was to turn the finale, previously a more or less light-hearted piece of contrast, into a centre of symphonic gravity. True, Beethoven was a great composer, and therefore not the best possible modern text-book candidate for the post of important historical figure; but at least he had this to be said in his favour: he wasn't out for perfection, and was therefore eligible as a revolutionary. He was one, no doubt, but it does not follow that the 'Eroica' was a beginning. Mozart, the ultra-perfectionist, never even seemed to qualify for the preliminary round of the Important Historical Figures' Cup; yet, to the unprejudiced ear, it is the 'Jupiter' that is a beginning – the most important work historically so far as the birth of the Symphony with a capital 'S' is concerned. Against the formal background of Mozart's own finale to the G major quartet, K. 387 (1782) and, it may be added, of one or two quartet finales of its dedicatee, Haydn, the finale of the 'Jupiter' redirected the symphonic forces of gravity once and for all and, in the process, re-created them. The afore-discussed G minor finale, is the last step in this development towards the complex last movement: though its total formal layout is fairly conservative, it is unusually tense emotionally and hence, as we have seen, explosively original, gravely meaning-laden, in some of the local structures it throws up.

Perhaps we are coming to understand a little better now why Mozart had to base this towering 'Jupiter' finale on a tag, i.e. the first four bars of Ex. 47 (with which compare Ex. 68 and Ex. 70 below – another anticipation of the finale theme, this time in the work itself: it is the theme of the twelve-bar middle section of the minuet's trio).

70

In view of the unprecedented complexity of the finale, its basic material had to be as simple, as 'well-known' or knowable as possible: the great creative mind's commitment to maximal clarity and comprehensibility is absolute.

The G major quartet's finale, the forerunner of the present edifice, is based on a similar tag (Ex. 71), but since that earlier structure is simpler, the tag can allow itself to be a little more elaborate:

71 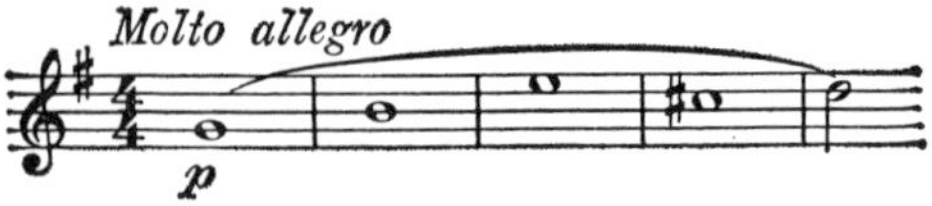

Now, together with the drastically enhanced substance of the 'Jupiter' finale, where fugue is composed against the background of sonata and, at the same time, sonata against the background of fugue, goes a proportionate simplification of the earlier movements – the opening one in particular. To an understanding ear unacquainted with the 'Jupiter', but familiar with the rest of Mozart's symphonic development, it would become increasingly clear in the course of the first movement, normally the most complex, that something new was going on – a re-balancing process which necessitated relative structural simplicity at the outset. There is very little here of the interpenetration of formal elements which we have observed in previous works; instead, Mozart retains a high level of meaning through small-scale operations, local tensions between foreground and background which are cast in surprisingly conventional larger-scale forms. In particular, a novel device makes its repeated and uniquely consistent appearance which, when it exceptionally occurs elsewhere in Mozart or Beethoven, is almost invariably attributed to the limitations produced by the range of the instrument in question – principally the piano, but also the 'cello. (A real composer of course always thinks in terms of such limitations.) I mean the melodic deflection which, since times serial, is known as octave transposition: a melodic note or series of notes is switched to the octave above or below. It is honestly incomprehensible to me that although this melodic procedure well-nigh dominates the first movement, it has never been noted: perhaps Ex. 72 apart, it is used too unobtrusively to penetrate the normal bad performance. Ex. 73(*c*) especi-

72

ally, is easily missed by the unattentive ear which expects Ex. 73(*b*), *and in fact gets it* in the flutes and oboes at the same time; the octave transposition is prepared by the preceding leap to the sixth (Ex. 73(*a*)) that replaces the expected conjunct motion to the second (D) – which however appears again simultaneously in the flutes. Ex. 74 is audible enough, but perhaps not striking enough to tired ears; yet the device,

73

74

which in this movement becomes a characteristic of melodic style, serves to express the finest differentiations in emotional content – nor is there even a theoretical possibility here that instrumental considerations are at the bottom of any of the deflections.

The slow movement (*Andante cantabile*), too, contents itself in the main with fairly local tensions between foreground and background. For example, a creative procedure is taken up which we last observed in some detail in the corresponding movement of the early G minor symphony, K. 183 (cf.

Ex. 36), in that the rhythmic foreground of the theme is metrically displaced against its background (dotted bar-lines), so that the background's upbeat becomes the foreground's main beat:

75

This tension is, of course, immeasurably more subtle than the early one: both foreground and background remain clearly defined and so fight each other openly; on the respective first beats of bars 2 and 4, for instance, the background seems to surrender unconditionally – only to resume the struggle with the *subito forte*s of the respective second beats.

To describe the miracle of the finale itself, as has so often been done, as the summing up of Mozart's symphonic art is demonstrably meaningless, even if we accept, as I don't, that the work was unconsciously intended as a symphonic testament: we have pointed out that with the partial exception of the great G major quartet, Mozart had never before done anything like it. The likelihood is that had he survived, he would have tried something like it again; for as opposed to Haydn and Beethoven, he gladly repeated his successful structural innovations. Perhaps that is why he is felt to be a conservative: he tended to conserve his own revolutions.

MOZART'S SYMPHONIES

Number of Symphony	*Key*	*Köchel Number*	*Number of Symphony*	*Key*	*Köchel Number*
1	E flat	16	18	F	130
2	B flat	17	19	E flat	132
3	E flat	18	20	D	133
4	D	19	21	A	134
5	B flat	22	—	D	141a
6	F	43	22	C	162
7	G	45a	23	D	181
—	B flat	45b	24	B flat	182
8	D	48	25	G minor	183
9	C	73	26	E flat	184
10	G	74	27	G	199
—	B flat	74a	28	C	200
—	F	75	29	A	201
—	F	76	30	D	202
—	D	81	31	D	297
11	D	84	32	G	318
—	D	95	33	B flat	319
—	C	96	34	C	338
—	D	97	35	D	385
—	F	98	36	C	425
12	G	110	37	G	444
13	F	112	38	D	504
14	A	114	39	E flat	543
15	G	124	40	G minor	550
16	C	128	41	C	551
17	G	129			

3
LUDWIG VAN BEETHOVEN
(1770–1827)

Basil Lam

*

> The intellect of man is forced to choose
> Perfection of the life, or of the work.
>
> W. B. YEATS

THEORIES about art are useless unless they are based on observation of actual works of art, and any theory of the symphony must begin with the truism that the greatest symphonies were composed by Beethoven. The sonata idea depends on the discovery that large structures may be erected on the basis of tonal relations, as may be seen by examining the earliest examples where themes are mere formulas and polyphony is banished, not through lack of skill, but because the harmonic tensions generated are incompatible with the setting out of designs based on simple triadic relations. Haydn's monothematic symphonic movements show conclusively that tonality, and not thematic contrast, is the essential principle. In its most elementary aspect this means the establishing of key with enough solidity to carry the weight of a whole structure and the discovery, in the mid-eighteenth century, of the means by which this could be done, was of such potentiality that composers rejected, without hesitation it seems, the wonderful and elaborate technique of the baroque masters, just as the forerunners of those same masters had rejected more than a century before them the no less perfected technique of the sixteenth century when they realized the possibilities of tonic and dominant and major and minor. Beethoven's work, then, is the consummation of a method founded neither on themes nor on polyphony, but on the generating force of basic harmonic tensions. The classical norm is the major scale and with it the

major triad. (Of Beethoven's three dozen symphony movements only six are in minor keys, and the proportion is smaller in Haydn and Mozart.) Erect triads on the degrees of the major scale and you find only two are major, those lying a fifth above and below the key-note respectively. Is there a simple chord, derived only from the notes of the scale, which is unique to that scale? Add the minor seventh to the major triad and you find only one specimen – that erected on the fifth degree. Such then is the basis of tonality: three major triads, the tonal centre lying between dominant and subdominant equidistant by five degrees above and below, and the unique dominant seventh. Introduce any other dominant seventh and you must change at least one note. Taking C major as centre to move to dominant or subdominant you must change F for F sharp and B for B flat – the two alternatives of medieval polyphony with its *musica ficta*.*
Notice that to confirm either dominant or subdominant with its own dominant seventh you destroy the rival triad – F sharp removes the perfect fifth F–C, B flat makes the G triad minor instead of major. However, the sharp leading note which is essential to classical tonality is, for the subdominant, available in the home key, E to F, whereas to provide it for the dominant you must change a note: F *sharp* to G. So the dominant represents a more definite move out of the main key than does the subdominant, and so it is to the dominant that sonata designs must make their first change of tonal direction.

These elementary facts form the basis of the classical symphony and may help to show that Beethoven's unmatched achievement is the consummation of an art founded not on contrivance but on nature (if I were not afraid of the metaphysicians, I would say on Nature). The drama and excitement, the lyrical inspiration, the human expressiveness of his music are subsumed in the mystery of order. When medieval musicians changed B to B

* To avoid the tritone F–B the B would become B flat or the F, F sharp according to context. These accidentals were not written down but would be added in performance according to the rules of *musica ficta*.

flat and F to F sharp they were preparing the way for him by exorcising the tritone 'diabolus in musica', the enemy of order and proportion.

Once we recognize these truths we can reject the superficialities of Beethoven's detractors who complain of his unadventurous harmony, or of his repetitions of phrases and metrical patterns. Such criticism fails to grasp the nature or intention of his art. As in all such matters, if we are to experience Beethoven's masterpieces according to their own intention, we must decide which of the elements separable by applying reason to works of the imagination are essential, and which are consequential or even irrelevant.

What then is a classical symphony? What does it state and in what terms? First, what it is not: it is not an abstract of autobiography. Let us study Beethoven the man with reverence and compassion, but to hope for understanding of the work from knowledge of the life is like reading a biography of Wren to aid one's appreciation of St Paul's. Who would turn to Vasari's *Lives* expecting to gain in understanding of design and colour in Michelangelo or Raphael? This is not to say that the lives of artists are without interest, but if an artist's work needs the commentary of biographical information to make it fully comprehensible the work is to that extent a failure. In music especially, the supreme works by which we measure other and lesser achievements exist in themselves, no longer expressive of their originators. It is best to consider Beethoven's music regardless of anecdote or contemporary comment, as though we were, like the fortunate students of Shakespeare, deprived by the chance of history of anything beyond the barest knowledge of the artist's personal life. Few attitudes to music are more diminishing of its power and integrity than that which treats it as a transcript of a composer's experience, in which the actual quantities of a human life are turned into a kind of spiritual algebra.

Beethoven was fiery and impetuous, impatient and careless of social conventions, and in earlier years full of self-confidence. It is easy therefore to regard him as a revolutionary composer, scornfully rejecting conventions in his eagerness to impress with his originality; unluckily his music refutes such an estimate, and emerges unaggressively from a tradition with which he never breaks. Although the sonatinas written when he was twelve show a characteristic vehemence, especially in minor keys, his first published compositions show no sign of the narcissistic eloquence of the inward-turning Romantic, bemused by the significance of his own feelings. It is not that an artist's experience is irrelevant to his activity – clearly certain elements can be isolated in a man's work and related to a common source in his life and thought, but such things provide the impulse to create – not the substance of what is created. *Lycidas* may be expounded as an expression of Milton's sentiments about poetry, death, the Church, rural scenery and the classics – the indestructibility of this particular masterpiece has been tried by just such an attitude. In the mid-twentieth century we are especially afflicted by the disintegrating effects of an unrestrained intellectualism and our refuge is the kind of Romantic art which expresses the torments and the erotic fantasies of those who against the advice of Carlyle are unable to accept the universe. A. E. Houseman said that the function of poetry was 'to harmonise the sorrows of the world'. This means neither to soften the impact of reality, nor to provide an imaginary world to escape into, but to assert the freedom of the creative will by producing order (harmony) out of the random and chaotic sequence of human feelings. This is the triumph of the achieved form which, in a striking phrase of E. M. Forster's, 'gives us temporary salvation' and it is because Beethoven has to compel more elements into meaningful and authentic correlatives to our apprehensions of life than any other composer that we must recognize that he transcends even the perfect art of Haydn and Mozart. To say this is to invite censure, for classical methods in criticism are out of favour, and the proper attitude to art is held to be an undiscriminating promiscuity directed to the enjoyment of as many things as possible, regardless of values. It seems that an imaginary line is drawn, above which it is illicit to remark that one thing is better than another; it is permissible to note that Mozart is greater than Dittersdorf and even to promote understanding by explaining why this is true. What arouses opposition is to maintain that even at the highest levels of art there is a hierarchy rather than an egalitarian republic.

Beethoven's First symphony is admittedly less remarkable in expression than many of his earlier compositions, nor can it match in subtlety or thematic invention the masterpieces of Haydn and Mozart. To explain this surprising fact it is commonly said that Beethoven approached the symphony with caution, advancing only circumspectly towards the more characteristic style already evident from his Opus 1 onwards. Such a view reveals its inadequacy once we consider what we know of the young Beethoven, both as man and artist. Yet it is true that much of the material in this first symphony is conventional, nor is the harmonic language adventurous. Why then did Beethoven not use any of his already achieved mastery of bold harmony and personal melodic invention? The first movement of the sonata in A, Op. 2 No. 2, moves with calm certainty of direction through modulations not containable within the 18th-century sonata. (C. P. E. Bach's paradoxes are the too-easy surprises of a style where anything may happen.) The F major quartet, Op. 18 No. 1, can absorb into a normal first-movement exposition progressions of a boldness and ambiguity for which the collector of such things could search in vain through the first and second symphonies. Evidently Beethoven was engaged in his first symphonic essay with matters excluding both the polyphony of which he was already the master (see the trio, Op. 9 *inter alia*) and the adventurous harmony which he had at his command from the beginning. Beside Mozart's 'Prague', Beethoven's First symphony appears almost primitive in these respects, with its plain harmony, its lack of thematic distinction, its avoidance of polyphony. Why did a composer of such superlative gifts both in technique and invention choose to deny himself the use of so many advantages when he turned to the symphony? If we can answer this question we shall have at least a starting-point for any attempt to understand Beethoven's conception of the symphony, and then to account rationally for his supremacy in it, a supremacy as beyond question as that of Shakespeare in drama.

Most difficulties here vanish if we regard the symphony as essentially harmonic architecture. As noted above, the so-called revolution of the mid eighteenth century was not to invent a formal scheme based on contrasted themes (these anyhow are often no more than the surface outline of the harmonic mass) but to realize the possibility of erecting formal structures held together by the tensions inherent in classical harmony. Modulation, in earlier styles, had consisted of a move away from a centre never really quitted, and regainable at any point with its concomitant main theme. The sense of measured movement in time on which all our music has so far depended, is produced, in preclassical styles, by rhythmic patterns and by contrast of theme and texture. In fugue, movement is a function of thematic structure and recurrence; there is no space in true fugal writing for harmonic patterns to be developed by the repetition of motifs, and the sense of movement depends on the spacing of the entries of subject and countersubject. The symphonic idea rests on the discovery that movement can be produced by harmony on a scale large enough to carry structure, and this seems to have demanded a suppression of thematic interest. In many early symphonies whole movements are virtually themeless, for everything is sacrificed to direct attention to key changes and to emphasize the energy of harmonic, not contrapuntal, rhythms. The triad, the dominant seventh, the elemental relations – tonic, subdominant and dominant – these are the foundation of symphony. These simplifications meant that individual parts in a composition no longer made continuous sense in themselves. Bach liked to play the viola because he was in the middle of the music. Such choice was no sacrifice in variety or interest as the movement of his harmony could sustain a mass of detail in polyphonic textures. Who would choose to play a middle part in an early symphony? There have been, and perhaps still are, fools who complain of Beethoven's 'uninteresting harmony', meaning that few striking and unusual chords or progressions can be found in his scores. Such things are incompatible with grand design in symphonic architecture.

With these considerations in mind, let us return to Beethoven's First symphony.

The introduction declares at once that we are to hear a working out of the basic harmonic facts. The symphony begins out of its key simply because if Beethoven chooses to state the proposition dominant seventh to tonic, he must, if he begins in the main key, either repeat the same progression or having stated his tonic, immediately move away from it, a tiresome paradox if attempted. So he begins with the subdominant and avoids the tonic in the second bar, to reach the dominant by the fourth bar. It is important to remember, as in the opening of the Ninth symphony, that the listener's awareness of events must not be affected by outside information. The first emphatic chord in this introduction is that of G, preceded by its dominant seventh and up to this point the music might be in F (if Beethoven were stating and then quitting the main key) or in G, approached from a strange angle (F and G are not really related). C major with the key note as bass is nowhere touched on until the *Allegro* begins.

This may seem an over-elaborate commentary on twelve bars, but the introduction has far more subtlety than can be accounted for merely by noting that the first bars are out of the key. The *Prometheus* overture begins with a far more 'interesting' progression, but the object here is merely to make a dramatic effect, which is far easier than to design a whole introduction in which everything directs attention towards the necessary event, the beginning, in plain C major, of the *Allegro*. The Fourth symphony is a larger treatment of the same plan, with a mysterious breadth which would not belong, even if the younger Beethoven could have attained it, to the Palladian clarity and logic of his first symphonic opening. The typically Mozartian profundity of the 'Prague' symphony's first bars – unisons stating D major leading to the first harmonized passage around B minor – was beyond the range of Beethoven at this time, and indeed would have seriously disturbed his mastery of formal architecture. How he was to transcend the perfection of Mozart can be discerned from the unprecedented breadth of harmonic span which makes the first theme of the *Allegro* the beginning of a new era. Compare it with the first phrases of a greater symphony also in C major.

(*a*) Beethoven No. 1

76

(*b*) Mozart No. 41

Beethoven's powers of composition are almost entirely concerned here with this extending of the stress-bearing capacity of the harmonic architecture. Once grasp this, with its consequent prospects of changing the whole classical style from the perfected comedy of manners (which is *not* superficial whether in Haydn or Mozart, or in Pope's couplets) to what is sometimes vaguely, if impressively described as cosmic drama, and we shall not commit the folly of carping at Beethoven's plainness of material in other respects. Obviously Mozart's exposition in the 'Jupiter' provides more original profundities than Beethoven's First symphony in all its movements (except for a passage in the minuet) but this is not the point. Beethoven may not have planned his symphonic activity on the assumption that he must make broader designs than the great predecessors whom he admired so deeply, but cathedrals are not made by drawing churches with the dimensions multiplied and the symmetries of eighteenth century themes could not be enlarged merely by increasing the distance between cadences. Symmetries there must be, or all is reduced to the anarchy of 'undisciplined squads of emotion', but the larger span of Beethoven's theme in his first symphonic *Allegro* cannot be reconciled with the exquisite balance of phrase and counterphrase found even in Mozart's broadest openings, like the impressive theme which strides up from the bass at the beginning of the C major quintet.

Mozart matches phrase with phrase and the expectation of symmetry is at once gratified, with no more detriment to the grandeur of the effect than the regular rhymes in Pope's loftiest eloquence as in the closing lines of the *Dunciad.* Beethoven enlarges our musical experience by extending our sense of time, the measure to which, as Leonardo remarked, all music must be subject. To account for this sense of space, of the liberation of consciousness from its own body-rhythm would require more understanding than we yet possess of the effects of music (or indeed of why it has any effect at all – another action-at-a-distance conundrum), but everyone feels that Beethoven did indeed enlarge the resources of the art, and this, if true, can relate only to the as yet mysterious physical-psychic operations of rhythmic and harmonic tensions. Perhaps there is a clue in some remarks of Bernard Berenson about an analogous element in visual art where space is less abstract. Of space-composition he writes 'This art comes into existence only when we get a sense of space, not as a void, as something merely negative, such as we customarily have, but, on the contrary, as something very positive and definite, able to confirm our consciousness of being, to heighten our feeling of vitality. Space composition is the art which humanizes the void.'* Foreground symmetries draw attention away from this void and music based on them attains perfection on the human scale – Beethoven's art brings the rhythms of Nature into relation with those of the human world and this, not technical 'advances' (who

* *Italian Painters of the Renaissance* (Phaidon edition).

could advance technically on Mozart?) is what takes him beyond other composers. Even the seemingly unambitious trio of the minuet in this first symphony shows the new concept of proportion in harmonic design. The section after the double bar begins with no fewer than eighteen bars in which nothing happens except an airy exchange between first violins and horns and clarinets playing around the dominant seventh. This is perhaps the first appearance of another revolutionary innovation, the use of almost empty spaces in harmonic architecture. A later and daring extension of this device is used in the rondo of the Fourth piano concerto.

In its finale the symphony introduces another Beethovenish idea also found later in the quartet, Op. 59, No. 1 (first movement). This concerns the handling of one of the crucial transitions in the sonata style – the return at the recapitulation. This will normally be another case of dominant preparation and the pleasure given by classical usage is that of punctuality, as Tovey put it. In this finale, Beethoven makes witty use, in Haydn's manner, of the seven-note scale which marks the beginning of the theme, including a dispute as to which way up it should go (inversion was not yet a serious matter). This point settled, the scale, in dialogue between treble and bass, proceeds through twelve bars of *ff*, after which the theme we have been awaiting during these highly dramatic events, slips in almost unobserved during four quiet bars of woodwind thirds descending to the tonic. This is perhaps the first notable instance of Beethoven's metaphysical wit, which we would do well not to describe as 'rough humour'.

With the Second symphony Beethoven's creative rejection of the expressive for the architectural becomes even more evident than in the first. Nothing could be more apparently formal than the theme which follows the vast introduction. Such severely constructional propositions appear in other works written about the same time (e.g. the Third piano concerto) and could be used only by a master of abstract design, for whom the emotion-carrying melodies of lesser minds are incompatible with the highest reaches of the shaping imagination. It is noteworthy that Schubert, nearest in power and depth to Beethoven of all who came after him, discovered in his greatest symphony the same potentiality in the plainest thematic outlines.

Beethoven's first movement here has for two of its main themes nothing more personal than a triad spaced out in various ways, an archetypal shape which was to serve for the Third, Sixth and Ninth symphonies. Evidently we shall never understand Beethoven if we suppose that he works with themes or subjects as the basis of structure. What is the theme in this *Allegro*?

It is a metrical unit like a 'foot' in prosody and a written out semiquaver turn; although the first four bars sound like the beginning of something of formal symmetry the counter-statement beginning on G expands so that the *forte* counterstatement comes after thirteen bars. This is no mere restatement of the opening; the first bar now moves up in three repetitions so that the flat seventh (C natural) is stressed and the D major so obviously, even sententiously presented in the opening theme is already threatened by the prospect of wider tonal exploration. (See bars 47–51.)

Notice how the turn comes at the third step after one beat, not three. This symphony will appear simple only to the casual and indolent listener who confuses Beethoven's plainness with the complacent formalism of lesser minds. Bar 51 has a shake or trill for three beats, presumably a casual detail to fill an empty bar. The main theme tends towards the tonic minor, but yields to an unmistakable transition to the dominant, a passage which incidentally gains in power by being written in two parts. (Bruckner knew what so many composers fail to grasp, that power is dissipated among a large number of parts, however impressive a passage may appear on paper.) Nothing could be easier for Beethoven than to 'flout convention' by arriving at an unrelated key for his second subject, but he is not interested in clever paradoxes and the theme is no less formal than the first. It is repeated with an added detail which is the measured trill mentioned above. This will soon be heard again. Before the exposition ends space has been found for a close sequence into which the movement's opening phrase is compressed, a device giving immense vitality to this section. Beethoven's marvellous technique – it is not less transcendental than his imaginative power – can be most clearly grasped when, as here, he is using the plainest of material. Now the development, asserting the tonic minor already touched on in the exposition, reveals a new rhythmic aspect of the first figure by omitting the first bar so that the pattern becomes 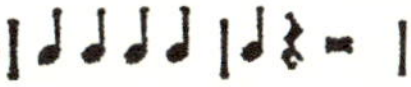 Over the first minor statement the shake comes three times with wonderful harmonic results. It is of course a decorated continuation of the inverted tonic pedal heard at the beginning. At this point the third bar of the theme is detached from its context, thus providing a new factor in the development. There are four crotchets in the second subject also and when these are similarly isolated the two main themes of the movement become associated, though when heard in full they are quite distinct. Thus the formal device of recapitulation gains in Beethoven's hands a new function by re-establishing the separateness of related themes.

In all sonata movements the beginning of the recapitulation is, or should be, a moment of structural significance: in the 'Eroica', Beethoven was to create unparalleled mystery and suspense. in this Second symphony he follows, with complete success, Mozart's admirable method of timing the return so that inevitability excludes drama. The Beethovenish touch here is that the development has reached F sharp minor (prominent in the harmonizing of the second subject) and the dominant of this key is heavily stressed. It is held unharmonized for two bars *pp* and then, by what was to become Beethoven's favourite device for modulation, an A is added and one further bar turns this third (A–C sharp) into the home dominant and we are back in D major with the first subject. This is condensed, with no counter-statement, and with the trill or shake augmented into quavers.

In classical schemes the end of the recapitulation offers an engaging problem to the composer. If the whole second group has been rested in the tonic then the *codetta* which closed the exposition will of course now be in the tonic, and Mozart can continue with his wonderful sense of proportion to let a movement make a natural conclusion at this point with practically no coda (e.g. symphony No. 38, first movement). Nevertheless, in most cases the dominant must not be too firmly established at the end of the exposition: if it is, the repeat will make the tonic sound like a subdominant. Consequently, in any fairly exact recapitulation the tonic will also not be altogether solid at its end, i.e. will be no more than tentatively a key in its own right. Beethoven reaches the expected D major in the movement under discussion, and by a quiet expansion of the last two bars of the exposition shows that more is to come (bar 303=bar 131). His problem is to retain the tonic as the underlying main key, yet to provide sufficient flexibility in harmonic design to make a climax in the coda. First he makes another cadence into the tonic followed by an emphatic C natural implying subdominant tendencies but nothing more remote, which here would have merely the surprising effect of the incongruous. Remember that a C natural made an unexpected appearance at the very beginning of the *Allegro* and you realize that this is an early example of the profound logic, far deeper than any thematic unity, which pervades Beethoven's vast yet ordered structure. From this flat seventh to the key-note arises a magnificent series of themeless chords over a rising chromatic bass (the momentum is so powerful that no theme is needed). The shock of these bold yet cogent progressions is such that there has to be a further reassertion of the tonic and of the characteristic figures of the main subject. Do not think Beethoven naïve when his last word on this subject is to march the four-note turn up the scale then down again. Such plain, even mundane utterances only confirm the truth of what has preceded them. Music had not known since Bach such a demonstration of the power of the harmonic imagination (see for example the G minor organ fantasia).

The *Larghetto* looks back to the eighteenth century and evokes with unassertive ease the world of Mozart's great serenades and divertimenti. Those who still suppose that Beethoven saw himself as a revolutionary innovator should ponder over the relaxed ease of this most agreeable of slow movements. It is in fact not really slow – the Handelian *Larghetto* is about the right pace – and there is an affinity with Handel's *bel canto* style traceable in the melodic span of the opening theme. The indolent breadth of this beginning is emphasized with the effrontery of an absolute master, by the repetition of both its sections by clarinet and bassoons (a Mozartian serenade colour) so that the complete statement occupies some forty bars of what is to be a design in full sonata form, with a far-ranging development. Evidently it is absurd to imagine that Beethoven first extended his forms with the 'Eroica'. This same development contains at least one modulation of remarkable power (bars 110–116).

The effect of this depends, of course, on its place in the harmonic structure of the piece and this is an aspect of Beethoven's mastery that is still inadequately recognized. 'Surprising' or 'daring' progressions are available to anyone. The works of C. P. E. Bach abound in harmony more 'advanced' than almost anything in Beethoven; the progression quoted occurs in Handel, another master of whose harmonic procedures complaint has been made, notably by Stravinsky. Surely it is plain that any musician who lets his fingers stray, no matter how idly, over the noisy keys will find in the twelve semitones of equal temperament plenty of ambiguous and enharmonic equivocations. The composer's art is to give necessity to the occurrence of some unexpected turn (surprise works only once) and in the passage mentioned Beethoven, having reached E minor, brusquely restores G sharp and then restates the whole process at a pitch leading to F, first major, then minor; evidently the enharmonic B flat = A sharp is only an incident in a much larger scheme passing through remote harmonic regions until, after a powerful climax in which a chord entered as D flat major is quitted within a single bar as C sharp major, we are confidently transported via the relative minor to the recapitulation.

In view of such music as this, let us not lapse into the still received opinion that Beethoven, after writing two promising symphonies, began to brood on Napoleon and found himself great with the 'Eroica'.

Mastery of proportion if complete must include the ability to contrast masses of various sizes and the scherzo of the Second symphony is notably concise and of chamber music clarity in texture; it could, with only slight changes, be transferred to the string quartet, which would be impossible with the 'Eroica' scherzo, where the genius of the piece depends on the latent power in the *pp* of numerous strings. Comparison here shows Beethoven's instrumental imagination in one of its almost infallible operations. It seems still necessary to restate the truth that his understanding of instruments (including the piano) was no less searching and creative than his inventiveness in theme and structure. Small though it is, this scherzo has left the world of the minuet and, as Tovey said, is the kind of piece Beethoven might have written at any time. His intellectual wit comes out in the trio. Having stated twice its miniature theme (in Haydn-ish scoring for oboes, bassoons and horns) he continues in the strings with a sententious insistence on F sharp and its major chord for no fewer than twelve bars – quite a long time in this minuscule piece. Two more bars of F sharp try to make yet another four-bar phrase but the whole wind-band with drums complete the phrase with an impatiently emphatic A natural, after which there is nothing for it but that the naïve little theme should resume complaisantly in its native D major, but robbed of its upbeat quavers, presumed lost beneath the unmannerly bluster of the wind.

With the finale Beethoven's power of movement attains its full maturity and range of contrast. He was still to broaden and deepen the classical style to its greatest possible attainment, but this magnificently confident piece must on no account be regarded as a transition to something bigger and better. The achievement here is to combine the concise wit of Haydn's great finales with the incomparable breadth and easy grandeur of so called 'second period' Beethoven. The first theme and the transition – this to appear to splendid effect in the coda – are compact ideas well inside the scope of eighteenth-century designs. Note how the transition theme is introduced by a formal flourish on the enhanced dominant only to begin in the tonic. It is with the 'second subject' that the composer's 'long majestic march' and 'energy divine' are manifest in a quiet dialogue sensitively scored for woodwind with a string accompaniment that throbs with unassertive power. You may if you take pleasure in such things, regard the woodwind theme as an inversion of the second subject in the first movement. The statement of this nobly-proportioned theme takes sixteen bars and the counterstatement, beginning in the dominant minor, expands with a sudden tonal inspiration into C major with one of Beethoven's characteristic phrases for the bassoon, an instrument especially honoured throughout his scores. As the movement opens in rondo style the Haydnish first subject shows that it too can expand and develop, and if what can develop is a theme we must confer that distinction on its first two notes the rhythm  (Similarly the theme of the first movement in the quartet, Op. 59 No. 3 is the pattern 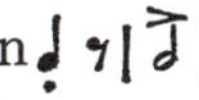) As mentioned earlier, the transition has its expansion reserved for the coda where its first two bars acquire a momentum sufficient to withstand two dramatic halts. The first of these is a *ff* dominant seventh followed by the dominant of B♭ minor, from which the onward march of the coda is all the more irresistible for being in an intense *pp*, wonderfully scored (notice the low horn pedal). Again the way to the final cadence (still a long way off) is barred, this time by two bars of a remote chord suggesting a foreign key which is thrust aside; we become aware of a new yet oddly familiar figure in the strings, which is of course the measured shake or trill so prominent in the first movement. When this twice makes a tremendous climax with the tonic flat seventh C natural (see first movement *passim*) the result is an harmonic unity far more satisfying than merely casual thematic resemblances which never seem to be structurally relevant even when their existence can be plausibly demonstrated. Now surely the end is in sight? But no – a mighty unison F sharp collapses yet again on to the dominant of B minor (remember the F sharp and its major chord in the trio) but this time the principal theme, after a momentary hesitation, combines with the accompaniment rhythm of the second subject: this produces a sense of irresistible impetus and when the last bars assert the quaver shake D–C *sharp* this figure seems by a marvel of design to have carried the weight of the whole splendid structure. No composition could more triumphantly express the separateness of, in T. S. Eliot's phrase, 'the man who suffers and the mind that creates'; it is a matter for wonder and gratitude that Beethoven should have overcome his personal tragedy in this heroic symphony, a monument of sanity and health. (The classical is healthy, the romantic ails, said Goethe.) Later composers have made so much of lesser griefs that they provoke the reversal of a phrase of Heine's – out of their little sorrows they have made large, or at any rate lengthy, songs.

Be this as it may, if you want dramatized autobiography in music, do not look to Beethoven. From this aspect it is especially regrettable that his Third symphony should have attracted an accumulation of irrelevant nonsense based on misguided attempts to relate it to its title and superscriptions, cancelled and retained. By all means let us render unto Napoleon the things that are his, but the greatest human hero would be unworthy of the 'Eroica' (in the European tradition, all valid heroes can have only one model) as we shall be if we let our experience of it be distorted by reflections on the French Revolution, a juvenile opera by Mozart, or anything apart from the miraculous structure in sound which is the 'Eroica' – perhaps more daring in its perfection and scope than any other composition.

To imagine that the first movement is based on a theme which is the major triad is like examining a large painting at a few inches distance in the vain hope of thus elucidating its form. Beethoven's drafts begin with the first six bars, always including the mysterious C sharp of which the consequences are so momentous. Expressive melody finds no place in these architectural studies, which appear to aim at balancing large-scale harmonic masses into a structure larger than anything previously attempted. Wren said that making the preliminary designs for St Paul's gave him a pleasure like that of composing music. That commentators should differ about the precise location of Beethoven's 'second subjects' should remind us of the inadequacy of such nomenclature, and engender a healthy nominalism. What was not usable here was any brief transition from tonic to dominant, though this basic archetypal relation pervades the whole vast design 'sustaining from above and lifting from beneath'. The point is made clear by a comparison with the theme that begins the *Allegro* in Mozart's E flat symphony. Here the admirable symmetry of the opening phrase spans the distance from tonic to dominant in four bars, whereas Beethoven's harmonic arch takes eight bars to reach the dominant (still not the phrase-end) and another four to return for the counterstatement which at once expands in broad sequences. A similar unprecedented breadth can be seen in the opening of the first 'Rasoumovsky' quartet, where, in a texture very like that of the 'Eroica', the music covers eighteen bars before the first tonic chord in root position. The still maturing genius of Schubert had not learnt this art from Beethoven when, in his nevertheless wonderful C major symphony, he made even broader spans by repeating short phrases. Now the 'Eroica' theme has a strong innate tendency towards one of these short-term tonic-to-dominant figures like this.

77

The scale of the design does not permit this to appear in the tonic, until, after more than six hundred bars, it can confirm in the final climax, the tonal centre of the whole movement. Earlier, near the beginning of the development, this same sequence is the means for extending the action of this abstract drama into harmonic spaces which would be mysterious were they not something greater – that is, sublimely rational in their inevitability. (The mystical is the antithesis of the mysterious.) Scarcely less remote is the consequence of the C sharp in the opening theme. This reappears in the development as the first step away from keys related to the tonic, the first phrase passing sequentially through C minor, C sharp minor to a *fortissimo* in D minor. At the beginning of the recapitulation this same note resolves by one of the greatest harmonic inspirations in all music, towards F major, unrelated and two degrees on the sharp side (i.e. dominant) of the tonic. To restore E flat as the true centre the theme is repeated in the key two degrees on the flat side (i.e. subdominant) which is, of course, D flat. The third consequence of this C sharp-cum-D flat is that it is the middle chord of the quite unclassical progression which announces the coda, E flat, D flat, C (all major triads). To describe even a few of the other harmonic wonders of this movement would go far beyond the scope of this study, but an appreciation of what Beethoven does in such matters basic to the symphonic idea can help us to value rightly the scope of his achievement, comparable with that of the architects who conceived the world's greatest buildings. If symphonies were buildings the listener would do well to quit some ambitious later examples lest they collapse on his head. Beethoven's harmonic scheme may be traced in the whole symphony, e.g. D flat is the major key used near the end of the Funeral March, after C major has been grandly present in the middle of the piece. C major enters with radiant effect in the finale, directly out of G minor (bar 257).

Little can be added to what has been said of the slow movement, though it may be worth remarking that this austere elegy is diatonic almost without qualification. Beethoven rejected at an early stage a chromatic descent from the main theme's appearance in G minor (this key, the mediant of the main tonic, is important in the 'Eroica' but little used in the first two symphonies). The stock resources for the expression of grief have been, since the Renaissance, chromaticism and 'poignant' suspensions and *appoggiatura.* Beethoven's majestic and impersonal sorrow is conveyed with none of these, and the utmost of desolation needs nothing more 'emotional' than the fugal episode, of which the rhythm is as unhurried yet as inevitable as the 'earth's diurnal course' of Wordsworth's brief elegy. It is interesting that Beethoven's sketches for another grave C minor piece (though the mood here is not tragic) similarly project then cancel the conventionally chromatic means of expressing mournful feelings. Beethoven must have thought such resources too facile.

> But I have that within that passes show;
> These but the trappings and the suits of woe.

To put it in different terms, the suspension and the *appoggiatura* both produce a momentary tension – relaxation of harmonic rhythm inimical to the processional grandeur of the long paragraphs in this march. It is because the whole piece has this fate-like and inexorable onward motion that the quiet cross-accents of the drums are so telling; when in the last bar this rhythm breaks down the effect is overwhelming.

> ...and I am re-begot
> Of absence, darkness, death; things which are not...

The C major section has the mood of Handel's Dead March in *Saul*: it is an error to regard it as a lightening of mood. Death is death, even in broad daylight.

What can follow a funeral march? Much ingenuity has been misapplied to account for the position of this movement in the 'Eroica'. If you are half in love with easeful death, like Tchaikovsky or Mahler, you can end your symphony with an elegiac slow movement, but Beethoven had no interest in the death-wish, which he had exorcised with the second symphony. The quiet energy of the scherzo's opening bars is the finest imaginable response to the end of the march. Violent contrasts come from the overheated sensibilities of the Romantics to whom the intensity of a sensation is all-important. What must, and does happen at this point, is the restoration of the normal human world without paradox or dramatic contrast: nothing else would ratify our trust in the composer's truth to the deepest experience. When the trio of horns take command in the middle section all sense of what Brahms disliked as '*Effekt*' has been averted by their appearance in the first movement. A lesser master would surely have reserved the use of this very rare orchestral colour for this moment. Incidentally, the adding of the third horn alters the whole balance of the classical orchestra by making available complete three-part harmony in a single colour. This is otherwise impossible with undivided strings. How to deepen and extend the range of tone for a symphony of unprecedented size and grandeur? Answer, add one horn. Technical mastery at this level can best be praised in the words of Sebastiano del Piombo when, in a letter to Michelangelo, he says of the latter's *Risen Christ*, 'The knees of this statue are worth the whole of Rome.'

In early sketches the scherzo was headed *Menuetto* and the three horns were to be prominent in the main section but not in the trio. Having decided to reverse this disposition of resources Beethoven wrote for the horns in the trio a theme obviously connected with that of the first movement, but not knowing the principle of thematic unity to be attributed to him a century later, he changed the theme to its present form with its melodic accent on the dominant. Indeed, unless we are to suppose that his untiring labours on the first three movements were based on a decision, already made, to use for the finale the *Prometheus* theme written before he began the 'Eroica', we must concede that the unity of at least this particular masterpiece rests on other than related themes. (Anyone who dislikes such inconvenient facts can always maintain that the 'Eroica' theme grew unconsciously from the *Prometheus* tune, so unconsciously that it nowhere appears in the sketch-books until he began to write the last movement.) As Nottebohm observed, the rarity of sketches for the finale suggests that Beethoven, when he planned this movement, must have determined to use not merely the theme, but much of the material and ground plan of the piano variations.

It is often said that after the introduction, which, like a prominent episode in the scherzo moves from the dominant of G minor to that of the main tonic, Beethoven makes first some variations on the bass of the theme. This is only in part true, for both melody and bass are essential elements in the scheme, the bass providing a subject for the two fugues and vanishing from the scene only when the tune itself is put in the bass for its last complete appearance. Beethoven's works abound in strokes of genius still unique because any attempt to follow them would betray its origin and be nullified. The 'Eroica' finale is without model or successor and becomes more, rather than less wonderful when we remember that it is extensively founded on an earlier composition not conceived for the orchestra. Variation form, in its eighteenth-century manifestations, was like the rondo, a means of relaxing the more rigorous demands made by the sonata-style on the listener's attention. The impression here of a giant rejoicing in his strength is happily reflected in the profusion of ideas which can be related to a pair of very normal themes. But there is no trace of decline in Beethoven's invention; a theme which can make a climax glorious enough to crown the largest of all classical symphonies simply by being played more slowly is without parallel and must be counted a notable feat of the creative imagination. There has been much speculation over Beethoven's choice of the *Prometheus* theme, but may not a factor be that, like Purcell's trumpet tunes, the whole melody can be played by the natural horn? How to turn the limitations of this instrument to positive gain may be discerned from the climax of the melody when its one high A natural becomes almost a *portamento* to the ringing B flat at the very peak of the phrase. It is often said that Beethoven's melodic style is instrumental, but behind it is the Italian vocal style which has been for centuries the source of true melodic invention; the human voices in the Ninth symphony make actual what is everywhere implicit in his music. His themes are things said but in a word-transcending speech of which we miss the unspecific meaning if we try to attach significance to them. Like all the greatest works of art, his compositions are not 'about' anything particular – they exist to realize in extension their own essence.

The 'Eroica' finale consists then of an introduction, a bass which furnishes a subject for two fugues, and a melody to which the bass is a foundation. Between the fugues come variations and the melody first reharmonized in a slow tempo so that the original bass is lost, then resolved into the Promethean theme. Only when the theme has been stated quietly, beginning in the austere tones of the wind (surely Stravinsky has admired this?) will Beethoven repeat it in the full orchestra. The wonderfully scored interlude between these two statements has an harmonic boldness which still sounds 'modern', plainly diatonic though it is.

How is Beethoven to end his heroic symphony? The architectural sense of design which is one element in his supremacy above all other composers in the larger instrumental forms has brought the climax of the theme in the right and necessary place, but this cannot be the end. We recognize that the solution is perfect only because we have it. Had the score remained unfinished from the end of the final statement of the Promethean theme no one could have guessed that Beethoven would recall the mood of the Funeral March in a passage of tragic intensity, in a key to remind us of the last episode in the March before its collapse into darkness. Now the drift of the harmony is towards a pedal bass in G minor, in which key the whole finale began. The descending scale of the introduction returns and all is accounted for, so that the tonic may be finally confirmed by fanfares on the Prometheus figure to complete this mighty design, the last two tonic chords balancing the two which began the first movement.

Until the Ninth, first and greatest of a new symphonic kind, Beethoven wrote nothing on the scale of the 'Eroica', apart from the no less spacious tonal architecture of the 'Emperor' concerto. All through his career he produced large expansive works interspersed with smaller lyrical compositions, for which he had a high regard, e.g. the sonata, Op. 78. The image of the frowning Titan which probably alienates as many musicians as it attracted a century ago, is based on a few larger masterpieces – the same may be said of Michelangelo's *Terribiltá*, which is far from the whole of his art.

The Fourth symphony is no less characteristic than the Third, though to undervalue it is a lesser failure in understanding than to praise it at the expense of the tremendous works on either side of it. In neo-classical circles it is fashionable to admire especially the 'even-number' symphonies while remaining silent about the others – a subtle though superficial mode of denigration. Such persons often present themselves as worshippers of Mozart, whom they would diminish by their regard, rather than Beethoven by their implied censure.

The Fourth symphony is no less a production of unlimited mastery than is the 'Eroica'. It is Beethoven's act of restitution for his youthful impatience with the aged Haydn. It has nothing of the strenuous brilliance of the Second, though there are obvious resemblances – the broad introduction, the lyrical slow movement followed by an epigrammatic scherzo, the Haydn-ish finale. The difference is between a work which announces a new era and one which, secure in its perfection, happily renders homage to the glories of one now past.

Whatever misleading inferences may be drawn from his social background and behaviour, Beethoven was, in his art, completely free from the easy-going bourgeois quality (*Gemütlichkeit*), with its concomitant sentimentality, which sometimes mars the marvellous works of Schubert and trails its way through the nineteenth century, infecting even the *Ring* (e.g. 'Brünnhilde as mortal woman') and, may we hope finally, pervades the music of Strauss and Mahler, redeemed partly in the latter by irony. Beethoven's comedies have the same intellectual keenness as Shakespeare's, a quality which enhances their untreacherous beauty and poetic depth: Holst said that music should have nothing to do with the domestic emotions. Evading the larger question of how music has anything to do with the emotions at all, may we note that Beethoven's introduction to the Fourth symphony is mysterious, because it does not reveal its dimensions nor insist upon its key, but it is not tragic, oppressed or in the least expressive of *Weltschmerz*. The tonic chord appears only once, after thirty bars, and then only in passing, as one step in a broad modulating sequence. There is no foreshadowing of later themes – a method used by Haydn – but the first sign of motion in this great vaulted space is a figure of detached chords, four in each bar. Everything is subdued in tone (the only *forte* moments are for *pizzicato* strings as in the First symphony's opening bars) until the brooding quiet is abruptly disturbed by the full orchestra with a typically and splendidly prosaic declaration via its dominant seventh that the key is B flat. The *Allegro* begins with no *accelerando* and with no theme until four bars have established the quick tempo. When the theme appears its quaver figure explains the mysterious process of the introduction. Later the uprushing semiquavers which at first seem to be no more than an introductory flourish reveal themselves as an essential feature of the movement's design. Such are the analysable elements in Beethoven's sonata compositions, those embodiments of an absolute intellectual mastery. This apparently minor detail gives unity to the development, its three versions being separated, and given each a precise and necessary function. Here are the three figures:

78

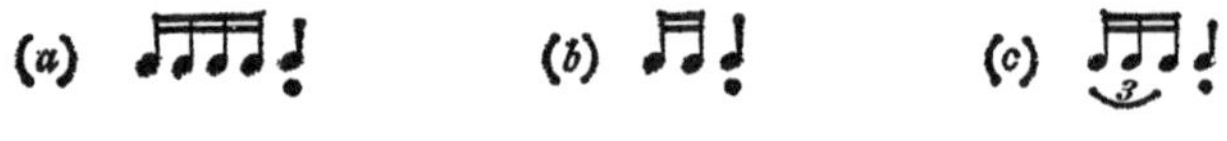

In the first movement of the 'Eroica' Beethoven had given a new dramatic meaning to the juncture at which the tonic is restored for the recapitulation. In the present work he extends the possibilities of stressing this feature beyond anything attempted in any other such instance, by building nearly half his development on suspense, both tonal and thematic. Before this a new *cantabile* makes a counterpoint to the first subject, which is then treated in the way normal to 'orthodox' developments until it is time for us to be reminded that Beethoven is a bard, not a gleeman, to borrow a valuable distinction from Robert Graves; tunes, however ravishing, are secondary to the assertion of order and the mystery of proportion and design. Everything now vanishes except for a rhythm and an harmonic motion suggested, not defined. Taken out of its context the beginning of this passage would be devoid of point or meaning; the harmonic direction changes imperceptibly to the ear when a reiterated D flat-B flat becomes C sharp-A sharp, to hover on the dominant of B major. But the timpani are never chromatic in the classical orchestra, so when a soft drum roll emerges in the middle of this chord it gives a faint suggestion that this wrong key is not quite solid: we know that the drum-roll must really be on the long-awaited tonic – B flat. Nevertheless a phrase from the first subject continues to move through the strings in B major – no equal-temperament woodwind is present – and when the chord of B flat is at last arrived at, it is in its second inversion, the drums alone establishing B flat as the bass. (Note the entry of double-basses in their deepest register.) When the recapitulation begins Beethoven repeats in the tonic the preliminary flourish which after the introduction had been in the dominant. This is exactly what is needed to reassert the tonic centre after the strange events just described. The dominant flourish is reserved for the coda which in its brevity recalls certain endings in Haydn, e.g. in trumpets and horns bars 483 *et. seq.*

For Beethoven the elemental relations of diatonic harmony are what the human form was to Michelangelo, an inexhaustible source of meaningful statements of which the form *is* the meaning. Every detail of what analysis misleadingly separates into harmony, melody, rhythm, orchestration and dynamics, must be exactly related to the total design: it is significant that Beethoven's sketches include dynamics, even sometimes in their earliest forms. Evidently his aural imagination included the stress-patterns of a whole sound-shape of which the realization depended on all elements making for a perfected whole. The coda of the movement under discussion provides an example. All that is happening is a tonic and dominant formula of the most ordinary kind, to be found in innumerable classical codas. The sudden *piano* in the *first* beat in bar 483 induces a drop in tension which induces a surge of power carrying the listener forward irresistibly to the final cadence. This *piano* is necessary as well as efficacious.

Berlioz tried to find words for his response to the *Adagio* (the only movement in the nine symphonies entirely in a very slow tempo, apart from the 'Eroica'), that 'undisturbèd song of pure concent' in which Milton might have recognized the embodiment of his most lofty conception of the power of music. The melody is the consummation of the *cantabile* style, non-sequential in pattern, no two bars being alike in rhythm, (another such theme is that which in a quicker tempo begins the A major 'cello sonata) and belonging to the same world as Handel's *Ombrai mai fù*. Though we can learn from such an inspiration that its creator must have been a man noble in heart and mind, this would be falsified if the music tried to communicate any such 'message' or indeed to do anything other than exist in its own beauty. The metrical beat which carried the 'linked sweetness long drawn out' of the theme appears in another of Beethoven's loftiest *Adagios*, that of the second 'Rasoumovsky' quartet, where, however, it is always 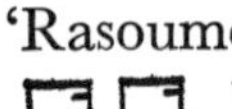. In the symphony it is sometimes this simple dotted rhythm, sometimes 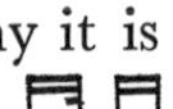. The use of these related but contrasted figures is a subtlety characteristic of the highest art; once done it is seen to be essential, though no analysis could predict its necessity. It could almost be said that the whole piece has for its main idea this pair of metrical figures, mostly quiet and serene but sometimes forceful, though never aggressive or violent, with trumpets and drums. At its return the main theme is decorated with such simple eloquence that it might almost be a reincarnation of the long-lost art of spontaneous embellishment. The *tutti* following this return shows harmonic affinities with the introduction; it moves over an impressively descending bass to the dominant of G flat where four bars of preparation for violins in two parts do nothing to slacken the pulse of the music even in this very slow tempo – a triumph for the classical sense of movement. The dialogue that ensues, between various orchestral groups and with the first four notes of the theme for subject, gives the dotted figure to high bassoon, then to 'cellos and basses, and finally as though discovering for the first time the true nature of this figure – to the timpani. The especial orchestral sound in this *Adagio*, as indeed for most of the symphony, comes from the prominence, in solo passages, of the single flute, the clarinet and the bassoon; the oboes are usually restricted to accompanying roles. When the second subject, that uniquely subtle theme for the clarinet, is recapitulated, the tonic key enables Beethoven to replace bassoons by horns, only one among many felicities in this score.

Although the scherzo begins like a quick minuet, in four bar phrases, the first moving to the dominant in orthodox manner, the harmonic range is unexpectedly wide and it is characteristic of this symphony that the second part should begin without explanation in D flat. In matters of rhythm, the regularity promised, or threatened, by the opening, is evaded when the first section turns out to be five phrases long, and the bar-rhythms of the second part are far too subtly proportioned for analysis in a brief study. When the principal idea returns it continues quite differently from the first time. In spite of the *Allegro vivace* indication the whole piece has a genial and leisurely expansiveness necessary to make contrast with the finale, which has the spirited rapid pace of the usual Beethoven scherzo. This relaxed breadth is most enjoyably expressed in the long tonic pedal in the trio (double-basses enter with grand effect at the climax of this long *crescendo*).

Studies like Ernest Newman's *The Unconscious Beethoven* combined with partly understood Freudian psychology have made it necessary to insist on the clarity and vigour of the composer's intellect. Every artist of the highest achievement must have a first-class mind and this finale of which the cheerful energy communicates directly with the feelings, has qualities also to give the keenest intellectual pleasure. To take a few instances. Under the semiquavers of the main theme a tonic pedal has the rhythmic pattern:

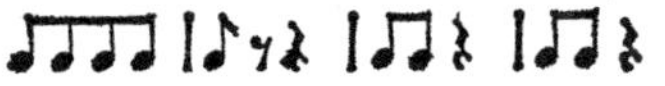

the outline of an important later rhythm. After the double bar the theme grows in sequences full of quiet energy, accompanied by the now complete form of this rhythm. These sequences reach a seventh over a B natural and when this foreign note has two bars of *ff* unison we are reminded of the tonal mystifications at a similar stage in the first movement. However this possibility is evaded by the unemphatic yet firm reassertion of a key much nearer home, the dominant of G minor, after which these sequences can explore a wide range of keys, while never straying too far to permit a punctual return. This having been made by the bassoon, with a slight feeling, however well played, of a *tour de force*, is repeated with derisive ease by unison strings; when the contrasting theme appears in the tonic, we realize, (if we have not done so already) that the chief strain of the first subject has been passed over (twenty bars have been lost *en route*) thus allowing for further developments in the coda, including a combination of two of the subject's principal figures. Before the end, Beethoven revives Haydn's comic device of spelling out slowly the notes of a quick theme. In this instance an especial point is made by including all the notes of the first four bars of the movement; when the mood of sentiment seems to be becoming dangerously prolonged, only a few *ff* chords are required to complete the movement. This, simple to describe, is a triumph of rhythmic organization no less than is the conclusion, at the opposite extreme, of the C minor symphony.

Although it surpasses immeasurably all its successors, Beethoven's Fifth symphony was the first of a new kind of symphony, which, because it expressed in musical terms the optimistic humanist philosophy, became almost the norm in the nineteenth century. In the works written on this general programme, emphasis is shifted from the first movement to the finale, where all contradictions of mood ('masculine' first subject, 'feminine' second subject, etc.) are resolved in a triumphant conclusion in the major key. This dangerously attractive scheme lends itself to rhetorical affirmations of shattering banality to which the fitting response can only be 'we might believe you if you didn't shout so much'. Beethoven's coda excludes such hollow apotheoses by refraining from any kind of thematic inflation.

It has often been remarked that his 'C minor mood' runs through his whole career. The works concerned are the piano trio, Op. 1 No. 3, the string trio, Op. 9 No. 3, the piano sonatas, Op. 13 and 111, the string quartet, Op. 18 No. 4, and the third piano concerto. Except for the last, all these show a characteristic vehemence, often produced, as Beethoven once remarked, by the use of the diminished seventh, which, for obvious reasons, will occur readily in the minor key. The music represented by this side of Beethoven's creative nature is that which has formed the popular image of the composer, not false but incomplete. In fact the C minor symphony was not at first planned to have its final shape at all. In the sketches a quite different finale was projected to have, so far as can be judged from the theme, a quality of pathos, rather like the last movement of the F minor quartet. The first four bars show an unmistakable affinity with Mozart's finale in his C minor concerto, which we know Beethoven to have deeply admired. Adherents of Schoenberg's absurd dogma that all themes in a work are related and derived from the first bars of a work, should try to explain the total lack of resemblance between this C minor theme and the actual finale of the symphony. On this point, it seems necessary to note that there is no connexion between the opening of the symphony and the repeated note figure of the scherzo. The real affinity is between the symphony and the Fourth piano concerto as the interspersed sketches reveal. The famous motive does occur in a scherzo, and in the right key, but Beethoven inconsiderately wrote this for the string quartet, Op. 74.

To add yet another analysis of the most written-about of all symphonies would be superfluous, but there is a general observation perhaps worth making in the context of this essay. In previous symphonies Beethoven had enlarged the scope of harmonic design by, so to speak, increasing the modulus. The unity of a composition depends, not on its themes in their melodic outline, but on the proportion of rhythmic-cum-harmonic units. With the Fifth, music's supreme architect constructs an edifice in yet another new way by making large sections divisible into phrase-units no longer than those of eighteenth-century styles. Study in this work the incidence of full closes in its various themes and this becomes evident. For example in the first twenty-two bars of the *Andante* there are six cadences in the tonic, yet the piece does not lack breadth. Again, the finale makes its grandly extensive phrases with a concentration of tonic and dominant motives of which the presence is stressed by literal repetitions. Throughout the symphony the ideas are of daring simplicity; Beethoven's achievement here is to draw on the tensile strength of the basic harmonic units while retaining his power to design structures no less extensive than in more complex works. This accounts, in one aspect, for the universal appeal of the symphony, which meets even Tolstoy's demand for an art comprehensible by all men. That Beethoven here addressed himself to the world in general is suggested by a feature of this symphony without parallel but for the Ninth. The return of the scherzo in the finale suggests an extra-musical motivation, for it is not to be explained in purely musical terms, though it would be a gross failure of response to question its validity. The reason must be a programme, though spiritual or metaphysical rather than in any sense material. The triumph of this movement, though enhanced by the largest orchestra yet heard in any symphonic work, is no easy affirmation. By recalling the opposite pole of experience in the scherzo – compare Goethe's Mephistopheles: '*Ich bin der Geist der stets verneint*'* – Beethoven adds the final conviction to his declaration of the victory of sanity and joy. It is not his fault that this mighty composition became a model for the inflated rhetoric of much later romanticism.

We may suppose that the idea of the 'Pastoral' symphony began, long before it was composed, when Beethoven recorded in his notebook a detail of observation (as Wordsworth might have done, had he been musical), relating to the sound of water in a stream. This was in 1803, at least three years before he began work on the symphony. His warning not to treat this work as pictorially descriptive has not always been observed; it is best considered in purely musical terms as one of the most beautiful and profoundly expressive of all classics. Indeed music quite obviously cannot represent visual images; the composer who wishes to illustrate is limited to the audible aspects of Nature. The 'Pastoral' symphony develops the harmonic structure of the Fifth but with even more simplified progressions drawn out over larger distances. If the reader will refer to the introductory remarks to this chapter he will find that the basic relations of tonic, dominant and subdominant there outlined form the material of practically the whole symphony, apart from the slow movement. Nowhere since the mid-seventeenth century has the common chord been so glorified as here.

Beethoven was no pantheist; 'The trees in the forest,' he exclaimed, 'worship God', but are accordingly not themselves God, and though he would surely have shared much of Wordsworth's attitude as expressed in the Tintern Abbey

* 'I am the spirit that endlessly denies.'

lines, he never shared the poet's faith in the natural order as an alternative to a lost faith in human perfectibility. What we know is that his response to Nature was too deep and intense to be called anything less than mystical, though one would not have dared to use the expression in his presence. The 'Pastoral' then does not express the Georgian poet's liking for country cottages. Nothing less than Milton's description of Eden in *Paradise Lost* will do for the slow movement, and the final rondo is as much a '*heilige Dankgesang*' as that of the A minor quartet. (Although it does not appear on the published score Beethoven wrote over this movement the German translation of the words *Gratias agimus tibi.**) His vision of the Golden Age is without Virgilian nostalgia and is as immediately present as the visible world in some often quoted lines from Traherne's *Centuries of Meditations*.†

How was Beethoven to mould a powerful and dramatic musical style to convey his deep untroubled happiness in the contemplation of country life? Evidently the tensions and reiterative rhythms typical of his middle period would be inappropriate to create – in a phrase of Cézanne's – a 'harmony parallel to Nature'. Yet by the force of such means the classical sense of movement is produced; all music is finally analysable in terms of the structural procedures capable of maintaining the measured subjugation of time. Considered from this aspect the 'Pastoral' reveals itself as a triumph of harmonic design. In the first movement, development must be a simplification, reversing its usual processes, if Beethoven is to justify his title 'Awakening of joyful feelings on arriving in the country'. Consequently the so-called development consists largely of a vast sequence of major chords, carrying repetitions of the second bar of the first subject. There is no other thematic interest and no

* 'We give Thee thanks.' The phrase goes on: *propter magnum gloriam tuam* – 'for thy great glory'.

† The passage beginning: 'The corn was orient and immortal wheat, which should never be reaped nor was ever sown. I thought it had stood from everlasting to everlasting.'

movement of the bass. The proportions are of primitive simplicity – twelve bars of B flat followed by twenty-four of D, then after a brief interlude the same process on G and E. This is perhaps the broadest expanse of harmony in all the classics, wonderfully scored (every detail contributes to the effect) and in its motion as unhurried as the sky-drift of summer clouds, yet as rapid as their shadows crossing a field. E major is remote from the tonic; the feeling of liberation Beethoven experienced in the fields and woods is perfectly reflected in this undramatic transporting of the listening mind to a key so far from the starting point – a harmony parallel with Nature – each change of chord revealing a further depth to the scene, and conferring Milton's 'sober certainty of waking bliss'.

To revert from the rhapsodic to the precise. E major approached from G is remote indeed, but the way back is so effortless that we realize, as flutes then lower strings proceed with bars 9–11 of the first subject, that we have remained near the centre all the time; it is the view that has changed. E is the dominant of A, which is the dominant of D, which is the dominant of G minor, and before long a held high C restores the home tonic, the recapitulation beginning unnoticed as the bass in a *diminuendo* falls from B flat to F. In the recapitulation and coda the triplet figure, originally a detail in the transition (bars 57–58) increasingly pervades the texture with a broad tranquillity which yet maintains the sense of a power like that of wind or tide. It is the material of the climax which, emotionally as well as formally, is the culmination of the whole movement. The final stroke of genius is the *piano* for the last two chords.

Although we know from the early sketch mentioned above that the two muted solo 'cellos represent the brook of the *Andante*'s title, no idea of the picturesque should impede our recognition that this is the most serenely beautiful slow movement ever written. It is not lyrical – the lyric is an expression of mood or feeling and to ascribe feeling to something so unmoved and expressionless would be a kind of pathetic fallacy. Beethoven's titles make it clear that whereas the human feeling about Nature is present in the first, third and last movements, this *Andante* and the Storm are in no way concerned with expressing human responses but with Nature itself. They show a landscape without figures.

Formally speaking we have here a movement in full sonata form on themes so free from tension that the hearer is drawn into a state of timeless contemplation. Berlioz found something like this in the slow movement of the 'Emperor' concerto where the piano broken-chord figures are indeed not unlike the accompanying textures here. As the harmonic motion is as slow as it can be without losing sense of direction rhythmic pulses are the means of producing movement, and these are to be found in the writing for horns and bassoons. Consider for instance how in the first theme the horns, having established a quaver pulse in the sixth bar, begin a pattern of displaced accents thus  etc. which, against the continuous semiquavers of the strings, produces a motion like that of the stream itself, still, yet always flowing. A new figure, evidently the transition, leads back to the first subject and it is only when the music moves towards the dominant that we recognize a new theme. As in the first movement all hint of drama in the sonata style is banished from this quiet world. A strange and wonderful harmonic idea comes when the expected cadence (at bar 33) brings instead a chord of A major, which followed each time by that of D minor, mysteriously adds depth by suggesting a modulation, which though possible is not made. After several returns it gives way to the main idea of the second subject in full dominant harmony. The development begins by introducing yet another theme – like the others in the simplest possible harmonic shape. When it has moved into G major the first subject begins to traverse a wider range of key, with orchestration marvellous in its simplicity. The theme is given to the oboe, with flute arpeggios and a string texture not found anywhere else – second violins and the solo 'cellos (still muted) an octave apart play thirds in opposite phase, with first violins and violas in octaves crossing those in a slower-moving pattern. Every time the theme itself is heard, the horns resume their syncopated figure. There seems no reason why something so perfectly felicitous should ever cease, and the harmony moves into E flat, led by the clarinet, and then, with the warmth of a Venetian landscape, into G flat. While the flute adds the violin and viola figure mentioned above, clarinet and bassoon begin what we realize to be the 'brook' figure, always until now heard in the lower part of the score, and the key moves to C flat. As in the first movement, when E major returned so effortlessly to F, Beethoven restores the tonic within two bars by the enharmonic change C flat = B major = dominant of E minor which can turn readily into G major = dominant of C minor which can lead easily to the dominant of B flat. Although it sounds laboured in description this passage is a *locus classicus* for the most imaginative use conceivable of simple chordal progressions. Note that when the dominant of the home key is reached Beethoven is in no hurry to resume the first subject and for several bars there is nothing but trills and descending figures in thirds for clarinets and bassoons. When the theme does return it begins in a single flute. One more stroke of harmonic design remains, which is to move again towards the subdominant E flat and to pause on the tonic chord for a moment (it sounds like a dominant at this point). Now come the famous and so absurdly misappreciated bird-songs. Do Beethoven's critics suppose him to have believed that nightingale, quail, and cuckoo would all sing together at a brook-side? This is, in regular normal phrases, a reminder that themes such as we have heard in this movement have their counterparts in the only true non-human music Nature possesses.

The avowedly picturesque scherzo has been adequately described. Its key relations are primitive but original. How many pieces start with a theme alternately in two major keys, tonic and submediant, with no linking modulation?* The contrasting section in 2/4 is reminiscent of Haydn. When the theme is repeated by violas and bassoons and

* But see the F major bagatelle of Op. 33.

later in all lower strings, the counterpoint with the violins is admirably astringent. It is remarkable that Stravinsky has not commented on this passage, so like certain things in his own music.

When the storm begins, its first menacing D flat breaks in upon a full chord of C major – it would be far less disturbing after a unison C. Just as the slow movement is not descriptive of a rural scene, so this piece evokes the power and majesty of a thunderstorm, not as something frightening because it is dangerous, but as the manifestation of a formidable and non-animate power. This cannot be done by simple-minded imitations of the actual sound of a storm, which could be far more effectively produced by a combination of thunder-sheet and wind-machine. Up to the storm only the idyllic aspects of Nature have been celebrated but Beethoven knew the hidden power behind these, and the 'song of thanksgiving after the storm' should be understood as including gratitude to the Creator for the storm itself, i.e. as evidence of the divine power, not as a prosaic relief that no one has been struck by lightning.

The storm then is a movement in Beethoven's most sublime style, beginning like a passage in the *Egmont* overture (in the same key); apart from the groups of four semiquavers in double basses against five in 'cellos, this piece is normal Beethoven in full grandeur with no programmatic idea needed to account for its development of a figure derived unmistakably from the 'brook' theme of the slow movement. (I do not understand the symbolism of this indubitable thematic relationship.*) Notice how sparingly the obvious resource of chromatic scales is used (this chromatic scale has one whole tone, so that the dissonances are accented), how the two trombones have only long notes, and especially how timpani, the most realistic substitute for thunder, play a role indistinguishable from that of a normal Beethoven climax until the storm declines, over a long pedal on the main dominant (open C string of 'cellos). An unambiguous symbolism may be found in the fact that the

* The effect of the storm on the water? (Ed.)

calm after the storm – *in*, not *on*, the dominant of F – is represented by a double augmentation of the storm figure.

By calling the final rondo 'Song of the Shepherds' Beethoven invites us to bring to our experience of this music all the associations evoked by such a title. To keep sheep was the occupation of those who lived in the Age of Gold. Apollo, the God of Song, tended the flocks of Admetus, and the shepherds of course received the news of the Nativity. Such are the images in the background of this finale, as innocently profound as Corelli's 'Christmas Concerto' or the pastoral music in *Acis and Galatea*. Schoenberg finely observed that the double pedal in the first bars resolves on a weak beat; this is like pushing off a boat and it sets up a motion as effortless as that of the stream in the *Andante*. Key relations are the simplest in the symphony; Beethoven is no more capable of an attitude of patronage to his peasants than is Shakespeare, but this is the countryman's thanksgiving and might well, without detriment to the mood of unaffected devoutness, include gratitude for a useful shower during the storm. The most moving bars, the emotional climax of the whole work are those marked, significantly, *sotto voce* near the end; the 6/5 chord at the peak of the phrase can only be called miraculous. The whole ethos of the work is epitomized in this quiet culmination; the *ff* passages are not functional climaxes at all, but breathe a majestic splendour (it is impossible to better Tovey's words) 'glorious as the fields refreshed by the rain'.

With the 'Pastoral' Beethoven completed his exploration of the structural potentialities in the so-called orthodox tonal system. As I have tried to demonstrate, he gave to each work a distinct architectural character, defined by its use of harmonic rhythms. In all the first six of his symphonies three out of four movements had been in the principal key and the introduction in the First, Second and Fourth had moved towards the tonic through various more or less remote tonal regions, none of which was defined enough to imply an alternative centre. Within the various movements the harmonic structure had been built around the tonic-dominant-subdominant relationship of classical tonality. With the Seventh symphony this principle is no longer valid; it is only because the work is so overwhelming in its power and conviction that this truth has been overlooked. A brief study of its tonal architecture reveals an extension of the classical scheme so far-reaching as to be virtually a new style which, however, remained without influence on composers' practice until modern times.* First, in the broadest outline, we have four movements respectively in tonic, tonic minor, flat submediant (middle section in the main subdominant), with the finale as the only other section based on the tonic. Even were the scheme within the movements orthodox in tonal structure this would be a plan of unprecedented boldness, and capable of sustaining an edifice only in the hands of a master who could establish his key centres with unshakeable firmness. Yet, in addition to this spreading out of harmonic stresses, Beethoven carried the same method into the movements themselves. Only over-familiarity with this composition can obscure the astonishing newness of its harmonic procedures. Four bars of a not very slow tempo (*poco sostenuto* – not *adagio*) establish a key but the fifth bar is the last inversion (the strongest position) of a dominant seventh *on* the tonic and therefore *in* the subdominant. The word of power is to repeat the A major chord but with G natural in the bass. The impact of this is sufficient to turn the harmony right down the cycle of fifths, so that the eighth bar is a chord of F major which is to be the key of the scherzo. The scarcely established A major of the opening is restored by the *pp* upward scale of that key

* It seems to me that Beethoven in this symphony strongly anticipates the so-called 'progressive tonality' of Nielsen; in the first movement F and C are notable foreigners to the tonic A, in the *Allegretto* they are more easily related to the prevailing A minor, and in the scherzo F major is strong enough to take over, the first change of key being to A, which is now itself so much a foreigner that it can behave only as the dominant of D, into which key the trio inevitably falls. After all this, only the greatest vehemence can restore A – hence the tremendous insistent energy of the finale. (Ed.)

over its dominant as pedal. The modulation here is typical of the work, the change of a single note rather than the 'normal' method mentioned early in this essay, of undermining one key by the 'enhanced' dominant in a new key. This scale grows mightily and pervades, with an energy like that of all the creative force in the world, the subsequent statements of the four minims that opened the symphony. This return, after two bars in the main tonic, makes the same powerful change to subdominant regions; the last inversion of yet another dominant seventh, that of C major, comes with such emphasis that A major is obliterated and the solidity of the new key is confirmed by a fresh theme in the woodwind, making this no mere introduction, however grand, but a movement with its own tonal and thematic ground-plan. The turn in this new subject (bar 23, 3rd beat) can reasonably be associated with the first bar of the *Vivace* theme. This develops quietly in an atmosphere not unlike that of the second subject in the first movement of the 'Pastoral'. When a formidable *crescendo* brings back the minims *ff*, beginning on the dominant of A major (the tonic!) the harmony yet again moves away towards B minor or is it E minor? (Note the force of the 6/4 chord – bar 38.) It is neither. C major is reached, with an indescribably grand effect produced by the scale figure beginning in the very lowest register of 'cellos and basses. This C major fades into the second subject as we may call it, now on the dominant of F major. This is the most consistently maintained centre indeed of the whole 'introduction' and continues now for eleven bars. The main tonic returns through nothing more solid than a unison E and the E major chord. As there is no other harmony in this passage leading to the *Vivace*, E might equally be a tonic itself.

Beethoven surely learned from Haydn to begin a movement with a lightly scored melody. The deep tonic pedal held by the horns in octaves shows the characteristic amplitude of texture in this symphony; among its other glories is the vastness of sound produced by the normal classical orchestra. Only a master who knew that the inessential diminishes effect would have written such music for an orchestra deliberately reduced in size from that of the Fifth and Sixth symphonies. It is easy to imagine how trombones would have damaged the texture by excess of emphasis. Following the principles of tonal construction found in all his symphonies, Beethoven establishes A major with especial solidity in the twice-stated first subject and long tonic pedals continue into the transition, when the first of the tremendous *crescendi* characteristic of this work is much intensified by the appearance in trumpets and drums of the figure 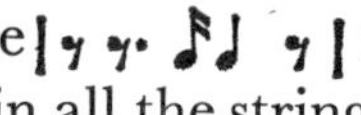 first heard under the flute in the first subject in all the strings. (It comes of course from the link between the introduction and the main *Vivace*.) The 'second subject' is not a contrasted theme, but in Haydn's way closely resembles the first, so that the effect is of a continuously developing texture dominated by the dactylic measure which is the real motive of the whole movement. Although this subject (which for orthodox minds begins at bar 130) starts properly enough in the dominant, it shifts into C major, returning to E through another *crescendo* dominated again by a cumulatively reiterated figure this time taken from the last bar of the transition, i.e. 129–30. Twice F major makes a sudden *pp* break in the twice emphasized *ff* on the now firmly established dominant. With the development, mainly based on the first subject, the genuinely revolutionary tonality of this work shows its logic and coherence. The first long passage, nearly thirty bars, on the beginning of the main theme is in C major (see the introduction) and the method of building remoter tonalities round the pillars of the central tonic and dominant is evident when this C major is abruptly superseded by the home dominant E major. Further modulations unequalled even by Beethoven in their range and unforseeable 'rightness' lead to a really immense climax on the chords D minor and A minor. As always, making creative use of technical limitations, Beethoven maintains through these bars an insistent A in the natural brass and drums, so that the home tonic rules these foreign harmonies. The recapitulation, delayed by two bars, begins in the full orchestra, so when the pause on the dominant arrives (bar 299 = bar 88) we expect from this composer that the counterstatement will be *piano*; but before the oboe begins the theme, with a heavenly lightness confirming that Apollo, and not Dionysus, governs all this power, another dominant seventh intervenes, so that the continuation is in the subdominant D major. The modulations effected by a single note (usually moving only a semitone) characteristic of this work reach the ultimate compression when the change of key to the coda is made by an *acciaccatura*(!):

79

The last note would be the leading note G sharp, if the grace note were a semitone lower. A flat, over a bass in the dactylic rhythm, changes to C followed by F (again, see the introduction) then almost for the first time in the movement, the dactyls vanish as an *ostinato* begins, derived from the first bar of the main theme; over it the intervals of the A major triad are spelled out against a dominant sustained through five octaves from flutes to horn pedal. Throughout this *crescendo* the writing is in two parts only and when all the woodwind and the horns affirm the main key with the first notes of the theme we realize that, after an unprecedented range of tonal contrast, the all sufficing climax is just to make this statement of A major.

The theme of the *Allegretto* which takes the place of a slow movement was, of course, originally intended for the C major quartet. It had a profound influence on Schubert, first appearing in an early sonata (conflated with the second movement of Beethoven's Op. 90) and later in the famous *Rosamunde entr'acte*, and thence in the A minor quartet and variations in B flat, in *Death and the Maiden*, *Die Sterne*, the A flat variations, and the C major symphony. No further comment could add to this evidence of the compelling quality which, from its first performance, everyone has felt in this movement. No familiarity should obscure our sense of the unique feature in design here – to construct a whole main section by cumulative repetition not of phrases, but of a complete theme. The elegiac mood is intensified when the music moves into the 'consolatory' tonic major. By the device, also unique, of beginning and ending the movement with the A minor chord in 6/4 position, Beethoven suggests that it is not wholly stable as a key as the implication is that it resolves on to a 5/3 of the dominant E major. Such is the prosaic account of what to the ear is full of poetic mystery.

The first bars of the scherzo began as the proposed opening of the first *Allegro*, and the important 𝅗𝅥 𝅘𝅥 figure appears, some time before the symphony was begun, in sketches for what was to become the *Namensfeier* overture. The double repetition of scherzo and trio was to have been used in the Fifth symphony, but the intense emotional force of the scherzo in that work presumably led Beethoven to modify his scheme. In the present work the repeats add weight to the D major tonality of the trio with its wonderful A inverted pedal, thus maintaining the tonic of the whole symphony as melodic, even when not tonal, centre. Furthermore, the complex bar rhythms of the scherzo reveal fresh aspects on repeated hearings, e.g. note how the accents of the main theme are reversed; see bars 334 *et seq*.

Blake, we are told, could frighten himself with a sense of the magnitude of natural forces by contemplating a knot in wood; the musician may do this with the finale of the Seventh symphony. The initial idea, interspersed with sketches for the earlier movements, was a symmetrical theme, remotely derivable from Mozart's E flat symphony, and of a type which could only have produced a movement lightweight in character. What the early attempts at a subject all have is the harmonic idea of starting on the dominant. That Beethoven's creative ordering of material was basically harmonic is further shown by a note against the transition theme 'this to go first into F sharp minor then into C sharp minor'. As Nottebohm observed, these modulations were used, but with different themes. The tonic is not established until the transition and is further removed from the centre by the tremendously insistent C sharp minor of the second subject, if such terminology still applies to the primitive quanta of energy used as themes in this piece. The clashing 'false relation' B natural in strings against the leading note of C sharp minor first heard in bar 144 is no casual detail and will be heard later. By the semitone shift found throughout the work – see the link between trio and scherzo (a unifying factor far more telling than thematic resemblances, which one should beware of exaggerating), the development moves into F – but over a C pedal, then, in accordance with the great overall harmonic scheme of the symphony, into C major, further emphasized by the repeats which also enhance the balance of the form enclosing these vast but controlled energies. From C major the tonality moves through F and returns towards the main dominant in a passage based entirely on two rhythmic figures, that of the first bars of the movement and the figure  (see bar 13 *et seq.*) With this material Beethoven develops the semitonal progression which rises in the bass, and then, with a stroke of genius outstanding even in a work where everything is so inspired, as soon as the home dominant is attained the tone diminishes and the flute has the first subject *pp* in B flat. The theme is given twice, in accordance with the overall scheme, and when the rising sixth recurs (bars 205 and 216) it is major, not minor (the rising semitone again) and by this change from B flat to B natural the bass has only to fall a semitone to begin the recapitulation on the home dominant. Everything is now centred on the tonic, to reaffirm A major as the central key of the symphony, until the end of the second subject. This, in the exposition, had carried its dotted rhythm through the dominant seventh of D, leading via a diminished seventh to C sharp minor. At this point in the recapitulation the home dominant replaces that of the subdominant in the expected way (bars 307–319 = bars 92–104) but instead of a diminished seventh comes a *dominant* seventh – of what? Of F major. It is impossible to write temperately of the mind which could conceive such fusions of reason and inspiration. (Note that this modulation is semitonal – C sharp to C natural and A to B flat.) The end of this passage brings a tonic pedal and the centre is restored. Now the exposition had ended the first time on the home dominant seventh, and after the repeat (to omit this is an unpardonable failure of comprehension) had begun the development in F by a semitonal modulation to its dominant seventh. The coda begins with the dominant seventh on the tonic, repeated four times with the utmost emphasis. It leads, not to the expected D major but to F sharp major, as dominant of B minor, and the coda begins in this supertonic region, the one related key not used so far anywhere in the work. Beethoven is now ready to build up, with the modest classical orchestra of his predecessors, a climax unsurpassed in its might and inevitability. The material is the first figure of the main theme in its extended form as originally heard in the exposition at bar 36 *et seq*. Then the first bar alone is developed over a bass moving in semitones (see bars 105–6) making a steady downward progression from the tonic to the dominant. When it reaches the latter, and not before, a *crescendo* begins over a one-bar *ostinato* E–D sharp. The clashing false relation mentioned earlier has gradually increased in prominence during the descent of the bass, and in the tonic as D natural against D sharp, increases the tremendous tension of the *crescendo*. Only a master of unlimited resource could thus dare to arouse anticipation, which is not disappointed when the scale figure from bars 32–36 becomes a rising sequence to culminate on a dominant seventh of the subdominant marked with Beethoven's very rare *fff*. We have still not heard the tonic chord with A as bass and even this climax must be repeated to fulfil the design. (The sudden *piano* at bar 435 is another inspired touch.) The momentum is now so overwhelming that even in the last bars the rhythmic emphasis can remain on the dominant chord until three bars from the uniquely abrupt end.

With this composition Beethoven lifted to the limits of human capacity his creations based on the principle of harnessing the power mysteriously inherent in tonality combined with cumulative rhythms. Of the latter a final word may be given to the metrical units which lie behind the themes and give a unifying character to the whole symphony. They are: first movement – ⌣ –, second movement – ⌣⌣, scherzo – ⌣, finale – ⌣⌣.

Between sketches for the *Allegretto* of the Seventh symphony Beethoven made the note '2te sinfonie D moll', but five years came to pass before he began work on the Ninth, and the Eighth followed after what was for him an astonishingly short interval, the autograph of the Seventh being dated '1812 13ten Mai', that of the Eighth 'Linz in Monath October 1812'. It is clear that from the first sketches the new work was to begin with a complete melodic theme, and so presumably to be on a small scale, at least in its opening movement; this as first performed was even more condensed than its final form, as Beethoven added 34 bars to the coda, which as we learn from the surviving drum part had ended *ff*. Although it is throughout an intellectual comedy in mood, the Eighth is really a new kind of symphony in form and proportion; everything leads to the finale which itself is a new form if we adopt an attitude of healthy nominalism to conventional nomenclature. The first example of the finale-symphony is of course Mozart's masterpiece; in the 'Jupiter', great as the other movements are, the finale is so transcendently inspired, that in retrospect, it is felt as the climax to which the rest of the work was leading. With his last symphony Beethoven was to find in the extension of this principle the solution to his life-long problem of finding a musical context for Schiller's Ode. In the meantime he surely wrote the Eighth in a mood of relaxation after the immense triumph of the Seventh. Nevertheless we would do well to avoid an attitude with any hint of patronage to this work of power and authority, not the less commanding for its good-humoured ease of manner.

When Beethoven, the master of movement, begins with a theme that stops at the fourth bar and is obviously going to be rounded off in the tonic at the twelfth, we might suppose that a dangerous humour was lurking. In this twelfth bar a new idea overlaps with the expected full close and at once the scale of the music expands enormously. The second subject, beginning in the submediant D before it settles agreeably into the respectable dominant, is far from symmetrical and its first idea is only the beginning of a grand and broad paragraph. The largeness of this movement appears in another aspect in the development, which is nearly as long as the exposition. The process which occupies its earlier stages seems to be a much enlarged version of an idea in the finale of Mozart's 'Prague' symphony. (N.B. This comparison is solely of size and not of value.) The passage in the 'Prague' is that beginning at bar 152; four bars of *tutti* on a sustained chord alternate with four more on the first subject in *p* woodwind. The movement is in a quick 2/4. Beethoven begins with the woodwind phrase from his first subject with the octave figure from the last bars of the exposition in the bass. The quiet section takes eight bars of 3/4, the *tutti* answer another four, and Beethoven, with a touch of effrontery, gives the whole process three times, the last *tutti* leading straight into a further development of the first subject beginning in the bass. With grand modulations this continues for what is surely the longest sustained *ff* in any classic, matched only by Brahms in his double concerto – fifty-four bars at the end of which the bass rises during two bars of *crescendo* and the recapitulation begins *fff* with the theme, not easy to hear in the bass, first given in a compressed form, i.e. bars 1–12 with bars 5–9 omitted. When, after the counterstatement, the transition begins in the bass, its power and breadth are enhanced by a splendid counterpoint in the violins. With the beginning of the coda we recognize the master of the Seventh. The recapitulation has ended firmly in the tonic and indeed the movement could, on eighteenth-century lines, end with a few bars of the main chord. Instead, the octave Fs in the rhythm ♩. ♪♫ continue in the bassoon then, changed to D flat, in the violas with the first subject in this unexpected key (but see the development bar 166 *et seq.*). Under sustained woodwind the five rising notes from the third and fourth bars of the first subject pass through the strings in a new rhythm  instead of  Although this passage is not long it gives an extraordinary grandeur to the movement, and Beethoven's harmonic architecture is recognizable in the role of D flat which as bass of one of his rare augmented sixths makes a second *fff* climax before the quiet casual-seeming end. This is deeper than mere wit in its power to remind us that all the vast energy latent in the 'little symphony in F' comes from a creative personality free from any trace of assertiveness or brutality. It also leads with complete appropriateness to the relaxed cheerfulness of the *Allegretto scherzando*, the lightest movement in all the symphonies, and one of the products of a complete maturity. Towards its end the subdominant is visited in the usual way, but the music remains overlong in this area, for, such is the compelling quality of Beethoven's harmonic-rhythmic structure, there can be only a few bars left before the final cadence. The problem is solved with magisterial abruptness and we must accept three bars as reaffirming the tonic, though our ears correctly inform us that this *coup de theâtre* is far from convincing. It is natural that Beethoven whose scherzo was modelled on Haydn's quick minuets (the variety of the minuets, e.g. in Haydn's quartets is one of the marvels of invention in music) should, when he himself wrote in minuet style, favour the slower kind. In the septet, the G major violin sonata Op. 30, the sonata Op. 54, the tempo is moderate, and the *Tempo di minuetto* in this symphony brings to this type an Olympian grandeur with no detriment to melodic inspiration. It is sad evidence of the blank incomprehension with which even great musicians sometimes responded to Beethoven that Mendelssohn should have conducted this noble piece in a quick tempo. Sketches for the trio give its theme accompanied by semiquavers – further proof that the pace is far from rapid. The difficult and highly original scoring of this section has been praised by Stravinsky. Only the desire for classification can obscure the fact that the finale is a new form in the symphonic style. You can, if you wish, call it a sonata-rondo with extended coda, but this is scarcely adequate as description of a movement in which the coda is nearly as long as all the remainder. The main theme is compact of subtleties. First, its half-bar beginning gives a witty ambiguity to the whole phrase structure. Then, the three-fold repetition of the figure:

twice over in the sequential opening, suggests a cross rhythm over the bar lines of three minim beats. As the end of the theme vanishes into a *ppp* over a dominant pedal our wish for overt logic and symmetry is hardly met by the explosive unison C sharp which appears from nowhere *ff* on a weak beat. However, it does incite the theme to repeat itself, this time in the full orchestra, and ending in the tonic, though it is still ten bars in length. Everything now proceeds with more regard to academic propriety, which would be mollified by an ostentatiously correct passage of dominant preparation, if the second subject did not insist on entering in A flat. This is soon put right with a beautiful quiet modulation and the theme is restated in the dominant. Note how the continuation in the woodwind conceals a Brahms-like transformation of the first subject in triplet crotchets (violins in thirds). Without the triplet rhythm the theme above in flutes and oboes (see bar 68) would not betray its origin. When this second subject continues vigorously in the dominant it becomes clear that the movement is in sonata form and presumably on a small scale. Unhappily for this prosaic though reassuring prospect, when the dominant seventh of C slips a semitone the brass and drums seize on this show of weakness to roar out the tonic and the future becomes far from clear. The violas, unaccustomed as they are to public speaking, timidly restate the tonic in cautious dialogue with the scarcely less intimidated violins. Gaining confidence they unite to begin the first subject; half way through the woodwind with 'cellos and basses break in and repeat a bar and a half. This is doubtless meant to be encouraging, but the thread of the discourse has been lost and the strings *pp* repeat the phrase yet again with irresistibly droll effect – 'but that is what we said'. The action soon grows in power and breadth, with a contrapuntal development based on the part of the theme represented by the rhythm  and the two rising notes at the beginning of the second subject, sometimes augmented. The subject from the first theme combines with itself in inverted form, and the whole section is of Bach-like authority and grandeur. It reaches a climax with the first theme in A major on a pedal E in octaves – a belated response to that anomalous C sharp in bar 17. The theme vanishes suddenly in mid-course leaving only the octaves on E. At the half bar this is annexed by bassoons and drums who raise it a semitone to F and the tonic being thus restored a recapitulation begins. The original statement with its three-part strings now has the drum and bassoon figure maintained as tonic pedal with *pp* horns and after four bars a new bass with sustained string parts as the melody moves into the woodwind. All this, so unobtrusive that the inattentive will pass over it, is a manifestation of supreme skill, the work of an absolute master at his ease, like Shakespeare writing the Falstaff episodes in the *Henry IV* plays.

The transition is expanded by nearly half its original length and when the second subject appears in D flat (replacing, with diplomatic correctness, the A flat of its first statement) we have another connexion with the intrusive C sharp though one that at the same time raises doubts as to its real identity. The recapitulation leads to another timid attempt to restart the whole process yet again, but we are in the subdominant now and no one has enough confidence in this key to go on with the theme. It breaks off accordingly, in comic dismay and there is a pause for reflection. A new theme begins in a phrase of great breadth accompanied by the  of the very first bar of the movement. (The theme in minims is in fact an augmented version of the descending figure in bars 5 and 6.) The minims become predominant and in a long development of formidable power, the first subject is represented by the augmented version of its triplet rhythm  for , as first heard in bar 18. At its climax the main theme enters in D major like one of Haydn's 'false returns'. The bass rises through two octaves with the same splendid release of energy as at the return in the first movement. The *ff* suddenly vanishes and we are left with the octaves on the tonic in bassoons and drums, exactly as before in bar 157, at what we were led to believe was the recapitulation of a small finale! This second recapitulation continues predictably enough until it reaches the anomalous C sharp. On its two previous appearances the intruder had been ignored, but confidence in orthodoxy has been sadly diminished by now and it is respectfully deferred to by the violas, by whom it is accepted as D flat. In response to this act of appeasement (perhaps he is only the flat submediant in disguise) the note insists that it is in fact C sharp, and when this too is accepted with servile readiness, C sharp is repeated three times and the first subject storms away in F sharp minor. Everyone joins in, except for the brass and drums, who cannot by their nature support such an unnatural key. However, the leading note of F sharp is E sharp, which is nearly F, and the moment this leading note appears in a register accessible to them, the trumpets and horns restore order and F major within a bar, and this coda, second recapitulation, last return of a rondo – whatever you choose to call it, continues in the tonic, reinforced as the true key by obstinate repetitions of the eighth bar of the theme. The second subject follows, in F major twice over (the only time in the movement it has not first been in a foreign key). The coda after narrowly avoiding a detour into the subdominant – always a lurking danger at this point in classical schemes – reiterates in high triumph the figures of the main theme and before the last *ff* Beethoven writes a wonderfully scored figuration of the tonic chord in thirds passing through all the wind over a long pedal. (Note how the drums enter, *pp*, with their octaves as a reminder that in these scores nothing is superfluous and nothing is wasted.)

Although Beethoven said that he experienced a feeling almost of dread when he realized that he was becoming possessed by the image of an imminent new creation, his symphonies, from the First to the Eighth had each been composed within a reasonably short period. Even the *Eroica* from first sketches to completion occupied no more than two years and nearly all the 'fundamental brain-work' on the score is recorded in a sketch book belonging to a single year, 1803. The creating of the greatest of all symphonies was not only laborious but impeded by uncertainties, doubts and the visitings of inadequate inspirations which had to be thrust aside. To follow in his notes and sketches Beethoven's struggle to realize his vision is to be reminded of Dante in the dark wood,

> una selva oscura
> che la diritta via era smarrita

though Beethoven was beyond the aid of any Virgil to lead him, having left the world in which his art had begun in cheerful confidence. It is necessary to remember that after the immense triumphs of his 'middle period' he wrote little for several years, and one can think only of 'a mind forever voyaging through strange seas of thought alone'.

In 1815, eight years before the symphony as a whole began to take shape, he wrote down the theme of the scherzo, headed 'Fugue', with the gloss 'End in slow time'. On the same page is a sketch for a symphony in B flat, bearing no resemblance to anything in the ninth, but with a memorandum containing the words: 'If manageable each instrument to come in one after another.' Two years later a different version of the fugue subject mentioned above appears as a projected fugue for string quintet. Then, still in 1817, Beethoven noted a group of themes and fragmentary motives, headed '*Zur Sinfonie in D*'. The sequence of these fragments only deepens the mystery which it reveals. The first is what became the transition (see bar 64), followed by the major version of the first subject as it appears in the coda, already here marked 1st horn with below – marked 2nd horn – the figure used, also in the bass, to link trio and scherzo in the Seventh symphony. Next comes a descending scale in demisemiquavers (see bar 34) then an open fifth *tremolando* in A and E with the same notes above in the rhythm . This hint of the great theme is followed by the theme itself, note for note as we know it. The way towards mystical understanding, we are told, is beset by devils with trivial and distracting visions, and Beethoven's last sketch in this group, headed '*letztes*' i.e. last (movement) is a hopelessly inadequate *alla tedesca* in D major. At this stage the movements were to be in the normal sequence, with the scherzo third, not second. Of the *Adagio* he knew that the key was to be B flat and that there must be four horns, two in the low B flat crook, but the theme is in 3/8 and is utterly remote from anything in the symphony. This rejected theme is followed by further sketches for the scherzo; at the end of these is the note: 'N.B. Here it must seem as though the trio is going to be in D, but it must unexpectedly go into B flat.' Now the trio in the symphony is, of course, in D, but it is an unforgettable moment in the *Adagio* when the key changes from B flat to D major and the *Andante* section begins.

Later in 1818, a memorandum shows that Beethoven contemplated two symphonies, one to include a chorus, either this or perhaps the whole work to be written in the ancient modes, but there is no hint of Schiller's Ode – the text was to be a *Te Deum*. Most important are two ideas – the voices were to enter in the finale or perhaps at first in the *Adagio* (which would therefore presumably immediately precede the finale) and, as alternative scheme, 'the *Adagio* to be repeated in a certain way in the finale, so that the voices then come in gradually'. At this time Beethoven was occupied with the *Hammerklavier* sonata and in the four years 1818–22 the symphony made no progress, even his creative energies being completely absorbed in the *Missa*

Solemnis and the last three piano sonatas. When he returned to his sketches in 1822 the first movement gained some fresh material including the idea of putting the first theme into B flat (as it was to appear in the *fugato*). Another theme, marked 'sinfonia 3rd movement' is of great interest, though not used in the work. This is a version in D flat, of the trio from the Second symphony!* There is still no trace of the slow movement, but the first line of Schiller's poem now appears for the first time with the opening phrase only of the great melody, which however gave place in a later sketch to a quite different idea in triple time, and then was abandoned for a time in favour of an instrumental finale on the theme later used for the A minor quartet. When, in 1823, the slow movement began to materialize, the second theme was written before the first. Here, as elsewhere in the work, Beethoven was vexed by trivial visions, and even as the sublime *alla Menuetto* as he called it reached its final shape it could not shake off an insignificant trio-like melody in its dominant key. The quest after the first theme was no less arduous and even after the essential idea of the opening had been written down, a second version started as a paraphrase of the slow movement from the 'Pathetic' sonata (Op. 13).

The finale was also being realized during the summer of 1823, with the theme gradually assuming its true shape. Even after the four bars which had been the whole of the first inspiration had been forged into the perfected simplicity of the great tune, Beethoven remained in doubt over the way to introduce the human voice; the instrumental recitatives were originally given words and these, written by the composer, aid our imperfect understanding of his intention, though were they not ultimately superfluous, he would surely not have suppressed them. The first words are: 'No this (illegible) reminds us of our despair.' Then the first movement is rejected as 'too unpleasing', the scherzo as 'no more acceptable, only more cheerful' and the *Adagio* with 'even this is too tender'. Eventually after wisely rejecting an appeal to 'join in the song of the immortal Schiller'

* Quoted in Sir George Grove's *Beethoven and his Nine Symphonies.*

Beethoven found the true answer and his own few words betray no inadequacy. Once the final problems of the ground-plan had been resolved the actual composition of the symphony seems to have taken not longer than in the case of its predecessors. What is unique here is the sequence of inspirations, unrelated to the order of events in the work itself, and the image we must have of an artist knowing somehow what he seeks long before he is aware of its nature. It is reported of Michelangelo that he described the process of sculpture, not as creating a form, but as releasing it from the stone.

Although it has had a number of successors, being dangerously imitable in its outward aspect, the beginning of the Ninth has no links with the past except possibly with the prelude to Haydn's *Creation.* All classical compositions begin with a theme, but here there is only darkness over the face of the deep, and a tonal centre which having no third, only a bare fifth is, by implication, major. It must be insisted that we have no foreknowledge, and A is the key at first. When, implementing the plan noted eight years earlier, the strings are gradually reinforced by other instruments, the real key is stated by the bassoons on an unaccented part of the bars (see the 'Pastoral', last movement and Schoenberg's comment) and this single falling fifth at once changes the insistent A from a putative tonic to an unambiguous dominant. Until the theme is heard in its mighty unison of the D minor triad, only bassoons and horns oppose the insistent A in the rest of the orchestra and in the whole statement from the first bare fifth to the cadence in the twenty-second bar there has been no harmony – only fifths and octaves. If we consider this opening as though encountering it for the first time, we shall realize that nothing remotely similar had ever been composed; or at any rate not for five centuries. Since the establishment of the interval of the third and thence of the triad at the end of the Middle Ages the expressive resources of harmony had grown through the centuries to be totally rejected by Beethoven for the beginning of his greatest symphony. Furthermore, as explained at the outset of this essay, the sense of movement in our music depends on the setting up of harmonic-cum-rhythmic tensions, and the Ninth begins in such a way that harmonic tension is abolished. Rhythm, which is all there is to give a sense of the scale of this opening at all, is itself so extended and unemphatic that for the first ten bars all we know is that whatever is to come is so vast that even its first phrase is too large to be grasped immediately. Notice how at the fifth bar one clarinet enters, at the ninth one oboe, then two, not four bars later, one flute. Two bars later comes the other flute and after *one* bar, both bassoons, with second oboe and clarinet on the last quaver of the same bar, followed on the second quaver of the next bar by the third and fourth horns. This sequence in bar units is 4, 4, 2, 2, 1, $\frac{3}{4}$, $\frac{1}{2}$. (See Beethoven's note of 1815 quoted above, p. 155.) Now even were this prelude to the great unison theme no more than introductory, it would still owe much of its power to these architectural proportions, but when the first *tutti* vanishes down a rapid scale of D minor and the opening bar returns now in D, we realize that this is no introduction but the first of a vast procession of themes. Because the opening was no mere cloudy introduction, its repetition in the tonic brings an expectation of an event at the fifteenth bar, when, the first time, the bassoons had changed the bass from A to D. Therefore when bassoons and horns fall not a fifth but a third, to B flat, the effect is all the more tremendous by depending not on surprise, but on the contrary factor – structural proportion. (A dramatic effect dependent merely on its unpredictability would not survive a second hearing.) The consequent restatement of the great theme in the submediant is unprecedented, for the D minor tonality is not really established until the motive  is detached from the complete subject and modulates from B flat back to the dominant of D minor. Even then it soon gives way to the second group, in B flat again. Only the master who had re-fashioned classical tonality in the Seventh symphony could dare thus to anticipate in the counterstatement of his first subject the key of the second. Everything Beethoven does in these compositions has a purpose beyond the immediate effect, and these events in the exposition have consequences later. Much has been made of the resemblance of the first phrases in the second group to the choral theme of the finale. This can mean nothing in the first movement, as the latter theme is still unknown and the whole point of the drama preceding its entry is the total rejection of the earlier movements, a sequence of events in which such a thematic reminiscence would be a distracting irrelevance. (What is not irrelevant is the motive of rising fourths in bars 80–81, which invert the characteristic falling interval of the opening and may be associated with the second subject of the scherzo (bar 93) and the fourths of the *Adagio*.) When Beethoven does not repeat an exposition we may expect some powerful stroke at the point where his intention becomes evident. There is a splendid example in the first Rasoumovsky quartet when the 'cello theme changes its direction with an emphatic G flat. In the Ninth, the exposition has ended with a triumphant fanfare-like unison stating the B flat triad in the rhythm . By the method of modulation now favoured by Beethoven, this B flat drops a semitone and we are returned to the first bar of the symphony with its measured *tremolo* on A and E. A new detail is the *pp* quaver interjections of trumpets and drum – this surely derived from Mozart's D minor concerto – but otherwise there is at first nothing to indicate that the repeat has not begun. Instead of some dramatic change of direction the music continues, *pp*, in the rhythm of the opening, the bass moving not to D but to F sharp, and this time with no *crescendo*. (Brahms remembered this in one of his deepest inventions, the middle section of the 'Tragic' overture.) Note that the bass moves, of course, on a weak beat. These bars with the almost motionless wind, the solemn trumpet and drum figure and the intense stillness of the whole texture, evokes the 'silence of infinite space' of Pascal's celebrated phrase. D major is here the dominant of G minor and in the latter key the main theme, still *pp*, is heard, in imitative dialogue. Everything is so vast and quiet that it is a shock when the *arpeggio* figure breaks out menacingly, this time on a diminished seventh. Under its impact the rhythm falters, but recovering returns to the G minor version of the main theme, which, now composed into a *cantabile* melody, humanizes the void. (The *pizzicato* double-basses, made commonplace in later ambitious scores, are especially telling here.) When the grimly energetic *fugato* begins, in the remote region of C minor, it might be supposed that it would develop towards a big climax and probably lead thence to the recapitulation. However, at the stage where this latter event might be expected (c. bar 253) there is a sudden drop in tone and we realize that the development is not complete. The theme, now in A minor, expands in a passage of Mozartian pathos and when the rising fourths of the second subject are heard in the bass in F major (bar 279) tragedy seems far away, except for the distant menace of trumpets and drum with their repeated A. (Note the silence of the double-basses in bars 287–297 and the sunlit clarity of the high flute.) Until the very instant when the D major return breaks in 'with flame of incandescent terror' only the change of one note, C to C sharp, has destroyed the F major tonality; This tonic major, with F sharp in the bass, is not new, for it had appeared at the beginning of the development, which, we now realize, has been an action compelled by a force like destiny in ancient drama to return to its beginning. As in the exposition, the bass moves to B flat but with a seventh added to the triad, to move with overwhelming power to the tonic minor. The theme itself now develops, with its second motive (bar 326) in the bass combined with the main figure above, then extended in dialogue over the delayed tonic pedal. As in the dungeon scene in *Fidelio*, there is a vast space between the double-basses and the rest of the orchestral texture. The D major–D minor pattern is continued with the second group and when the wonderful change to the flat supertonic brings, for the C flat of the exposition, E flat major here, we grasp the relevance of the dominant seventh of this key at bar 313. With classical logic the recapitulation ends on the fanfare-like assertion of the D minor chord. What possible dramatic stroke could begin a coda to follow the most tremendous recapitulation ever written? Perhaps the tonality might change abruptly in an unexpected direction and there begin some vast *crescendo?* Beethoven continues quietly in D minor with the *cantabile* version of the main theme over a *pizzicato* bass, a combination already heard in bars 198 *et seq.* but having expanded on a scale large enough for the greatest scenes in Wagner. We shall understand what supreme mastery means in musical architecture if we observe that such things in Wagner take their place in a design lasting an hour or more, whereas Beethoven's whole first movement occupies only some fifteen minutes. There is no change of key and the expected *crescendo* has a transitional quality that excludes any sense that the end is near. Indeed it declines and then builds again only to fade into a sustained A in all the strings. The serene D major version of the theme given to the horn is, with its accompanying figure, what Beethoven noted in his sketchbook in 1817 – the form latent in the stone. As in the first bars of the development this tonic major is shadowed by the menace of the low trumpets with the drum repeating the dominant and this grows to a steady tread as the minor third reappears and massed strings in octaves overrule the dialogue between single wind instruments. Twice the *crescendo* rises and declines, the second time into the pathetic hesitating phrase already heard twice in the development, in such differing contexts that we cannot anticipate its consequence here. When it is repeated and the last phase begins, the chromatic bass has a tragic power enhanced by the extreme rarity of such a procedure in Beethoven's diatonic art, and by the fact that the new figure above it is a plain tonic and dominant formula of extreme severity. It is characteristic of the whole movement that the three-part writing should coalesce into octaves at the end.

By placing the scherzo second in the sequence of movements, Beethoven created the same problem as in the 'Eroica'; how to make an energetic piece follow a tragic coda without falsifying the antecedent statement by over-dramatic contrast. In both symphonies the scherzo begins *pp* with intensely controlled rhythmic power, but as the previous movement in the Ninth had ended as decisively as could be imagined some link was essential to maintain the dramatic truth which in this composer is never tinged with melodrama. Beethoven's unforgettable solution is the terrible grotesque opening in an eight-bar phrase of which five bars are silent. Now he can write the fugue in D minor first projected in 1815. A strict fugue for orchestra is an aesthetic impossibility and an infallible instinct led Beethoven to Bach's method in all such polyphony, so the wind supply a continuo to the fugal exposition. The fifth entry, given to independent double-basses, settles on a long dominant pedal, thus establishing a harmonic-structural affinity with the first movement. When the counterstatement in the full orchestra proceeds to a sonata-like change of key it becomes clear that this scherzo is indeed in sonata form, with a second subject in the utterly unexpected key of C major. This too is on a tonic pedal, the furious octave Cs of all the strings obscuring, as perhaps was intended, the woodwind theme, which has a *Walpurgisnacht* atmosphere in its coarse geniality. This passage is often quoted as one of Beethoven's 'miscalculations' in scoring. It is hard to write temperately of such crude failures in understanding, in face of the deaf composer's countless marvels of orchestral imagination everywhere manifest in his later works. This exposition ends quietly with a return to D minor followed by the silent bars of the opening. It is difficult to account for the disturbing and tragic effect of these silences, the second coming after a hesitant move to E flat. The octave figure carries a series of modulations, dark with mystery, leading to B minor. (Note the savage emphasis of the *ff* horns in bar 173.) Throughout the exposition, four-bar rhythm has been maintained with a regularity not found in other of Beethoven's scherzo movements. This feature gives extraordinary force to the long section in three-bar rhythm which follows, moving, always *pp*, from B minor to the dominant of A minor, whence it is thrust by a single bar from the drums with their octave figure in F. Any other composer would surely have used these octaves for the *tutti* at bar 57. Another stroke of genius at its most clairvoyant is the vanishing of the drums in bar 208. Having brusquely taken over the three-bar phrases by answering the first bar four times over, they disappear in the *second* bar of the fifth phrase, with the result that in the rest of this section the pattern is altered with horns marking the second of the three bars with the octave figure. At the almost imperceptible return of four-bar rhythm, although everything remains *pp*, this figure begins to appear more frequently until with the wonderful modulation into E flat, it appears in every bar. The reappearance of the drums in bar 248 is another inspired detail. The return of the first subject is much enhanced by the static effect of the last passage in the development where in eight bars the only rhythmic motion is the octave figure. The transition to the second subject is lengthened by four bars, also dominated by the drum octaves. As everywhere in these works, meaning accumulates as the music develops. Who could foresee the consequences of that startling fifth bar? The second subject comes first in the tonic major then in minor, reflecting the tonal structure of the first movement. If the long repeat which Beethoven once restored is observed, the listener feels himself caught up in a dynamic process from which escape seems impossible until it happens with the complete contrast of the trio, which might be read as a musical interpretation of Wordsworth's *Immortality Ode*, had Beethoven known the poem, though it has insights deeper than the poet's. For the first, but not, of course, for the last time in the tonal structure of the work, D major is associated with a mood of complete happiness, in themes and textures that everywhere exemplify the profound difference between the childlike and the childish. Now that the exploitation of nostalgia for childhood has become one of the most discreditable features of twentieth-century art, it is fatally easy to misunderstand Beethoven's music in this part of the Ninth symphony. First we must banish the fallacy mocked by Johnson with

'Who drives fat oxen must himself be fat'. Although the piece radiates an unequivocal innocence Beethoven's attitude is neither naïve nor patronizing. The mood may be dove-like but the wisdom of an infinitely experienced serpent reveals itself in the handling of the archetypal theme. For a single instance note the heavenly length of the oboe solo – after the incomplete bassoon entry of the theme in bar 454 – with its brief excursion into F major. There is surely a profound significance in the increasing prominence of long pedals as the trio progresses. These, in the first movement, were bearers of tragedy, but here shine through the music like the unobscured sunlight of the 'Pastoral' symphony's first and last movements. Beethoven once thought of introducing three trombones at the end of the first movement in his as yet unwritten symphony. Once he had decided on the choral finale tradition would prescribe the appearance of the trombones with the voices, where they are reserved for a moment of especial gravity. So their entry in the last section of this trio, far from being adventitious, completes the transparent depth of sound which gives to the music here a quality worthy of Milton's words (themselves evocative of the end of Dante's *Paradiso*)

> . . . the Fair music that all creatures made
> To their great Lord, whose love their motion sway'd
> In perfect diapason, whilst they stood
> In first obedience and their state of good.

Nothing could be more moving than the abrupt fading of this vision before the ceremony of innocence is drowned with the return of the scherzo. To reject a convention is far easier than to give it a new meaning, and the formal repeat here would be necessary in the context of the whole work, even had it never previously existed.

The D major theme of the slow movement came to Beethoven before the main subject, and he was seemingly in doubt whether to use it, judging by his nephew's touching entry in a conversation book of 1823: 'I am so glad that you have brought in the beautiful *Andante*.' What can be abstracted as form in all great music is the embodiment of an idea (in the Platonic sense) and the Ninth, in many ways a larger and deeper restatement of what Beethoven had said in the C minor, follows in its slow movement the scheme of two strongly contrasted themes. However, in none of his earlier symphonies had the slow movement been personal in expression. Here the two first bars express a deep pathos such as Beethoven rarely admitted to his music. It is recognizable in the *Adagio* of the first 'Rasoumovsky' quartet (entitled in the manuscript 'A willow on my brother's grave'), the middle movement of the F minor quartet and of course in the *Cavatina* of the quartet, Op. 130. Such chromatic inflexions as that in the tenth bar of the symphony's *Adagio* are as infrequent as the no less expressive suspensions in the preceding phrase. Over the whole melody, with its wind echoes at the end of each phrase, Beethoven could have written the words he put at the beginning of the *Missa Solemnis*, '*Von Herze – Möge es zu Herzen gehen.*' Not only is the change to D major effected without its dominant but the whole strain is on rather than in the key, for its own dominant is nowhere stressed. (In the sketches this happened later, in the section Beethoven excised before completing the plan of the movement.) Only Gluck at his most inspired approaches the perfect simplicity of this *Menuetto*, as it was called when first written down. (The minuet which frames the flute solo in the Elysian Fields scene of *Orfeo* would provide an example.) The whole episode, it should be noted, is virtually on a pedal A – this note being held throughout by the 'cellos – so that when the main theme returns in a simple variation, its key is restored without emphasis. When at the close of the variation the modulation is extended by a whole bar, the change from D major to the G major triad on the last beat sounds as fresh as if this were the first day of the musical creation. The restatement of the episode in G means that we have the pedal now on D, the tonic of the whole symphony. Although the melody is unaltered new details make it ever more beautiful without detracting from its simplicity, e.g. the violas in bars 71–72.

As it approaches its end in a texture of celestial lightness, with *pizzicato* strings and the inexplicably felicitous horn octaves we expect a notable modulation. Expectation is adequately answered when the dominant seventh of G merges into the chord of E flat, and in this new key the main theme begins what appears to be another variation but is in fact a mysterious episode moving into C flat – a tonality not heard since the first movement (bar 108 *et seq.*).

That the occasional can influence the creation of the permanent is shown by the circumstances in which Beethoven wrote the horn solo in this episode, which must have struck its first hearers as the realization of an impossibility. The simple explanation was that one player of the instrument in Vienna had a primitive example of the valve-horn: hence the allocation of this elaborate solo to the fourth horn. Before this episode, so remote yet intimate in expression, returns to B flat and another decorative variation, the horn has what is virtually a *cadenza*. In his youth Beethoven is said to have enjoyed his ability to move his listeners to tears while remaining detached from the emotion he aroused. Something of the kind may be recognized here, for even in the middle of this *Adagio*, there is room for the element of play, in a moment of inspired levity, which Mozart surely would have understood and, had he been musical, the ageing Yeats, who could write 'Hamlet and Lear are gay.' It is not the intention of these variations to move away from the theme, so the melody is retained, unchanged but for its triple metre, in the wind. The architectural frame of the movement has rested on a series of great modulations, so when the variation reaches the dominant seventh of its twice interrupted final cadence, expectation is high, especially as these bars are enhanced by the famous triplet figure of the much-honoured fourth horn. As before, expectation is not cheated, and the key changes, with ritual solemnity (trumpet fanfares) to the subdominant, only to move back to the slowly descending fourths of the theme in darkening harmonies. It is a great moment, even in this work where everything is so inspired, when high violins, answered by flute and oboe in octaves, and borne up by the *pizzicato* of all the other strings, banish the shadow, as it were, with a phrase of the theme, significantly marked *dolce*. As the theme continues in the bass (bassoons and horns), with a countermelody in flute and oboe, the mood is that of the *Benedictus* in the Mass. (Note how the violins simplify their *fioriture* into a measured trill, echoing the triplet figure of the horn a few bars earlier.) By the characteristic change of A to A flat the solemn fanfare returns, again in E flat with its *Magic Flute* associations and, as before, turns back towards the tonic of the movement. To make such grand gestures is easy and perilous unless you can match the magnitude of the expectation they arouse. The descent into D flat, powerful even in isolation, gathers to itself the accumulated weight of all the previous modulations in this broad design, so that in retrospect – the only sense in which musical architecture can work – the whole key structure seems to lead to this revelation. From the great darkness of the next bars the return to daylight is yet again made by a single semitone, D flat to D natural. (Note how a fragment of the fanfare rhythm echoes through bars 133–6.) All now is epilogue, its first part marked with the '*cantabile*' reserved for passages of an inwardness the antithesis of '*espressivo*' and floating over a *pizzicato* accompaniment. In this music where everything is accounted for, the pathetic harmony of the two opening bars is heard again to make a shadow soon dissolved in the calm light of the close.

One problem is inherent in the classical symphonic scheme. Each movement must be a complete design in itself, yet must also imply, for all the finality of its conclusion that something is to follow. The finale of a large symphony must resolve in its coda the energies accumulated through the whole work and when these are of unusual intensity, as in the Fifth, Beethoven is likely to incur the disfavour of those who cannot grasp the necessity of reiterating 'mere chords' until the thing is brought to a halt. To manage the transition from a notably elevated slow movement to a healthily energetic finale he will sometimes evade the concluding phrases of his coda and proceed to administer the pleasurable shock of a plunge into cold water with a tactlessly contrasted rondo theme. Examples doubtless painful to over-refined sensibilities are the violin concerto, the 'Emperor' and the great B flat trio.

No such drastic solution will do here; to attempt it would be to make the error fatal to Gluck's *Orfeo*, of which the happy ending betrays so lamentably the great music of its earlier scenes. It is not surprising, therefore, that the transition from the *Adagio* to a vocal finale cost Beethoven much effort before he found the true answer. It must be said dogmatically that the last movement is the necessary outcome of the rest of the symphony. Those who reject these choral variations must deny the validity of the whole work which in tonality, and in the sequence of ideas and emotions, leads to its destined conclusion. If Beethoven had used the tragic theme we know in the A minor quartet he would have written another and different symphony. When an artist has shown in a series of works universally acknowledged that he possesses structural mastery matching the depth of his imagination and all other elements necessary to greatness, it is unlikely that he will blunder in a masterpiece of his fullest maturity. Those people who reject the finale of the Ninth would do well to examine, in an attitude of decent humility, the possible causes, in their own experience and emotional life, of such a failure in comprehension. Beethoven is not infallible, but he is no more capable than Shakespeare of faking an emotion or contriving effects with insufficient or flimsy causes. If you cherish art as an immaculate substitute for life you are bound to be offended by an artist who having created a work of unsurpassed perfection and depth, says, in effect: 'This is not enough; art must not set itself above the humanity which it serves.'

The *Adagio* has ended in untroubled calm, with no tensions unresolved, and although what follows has been the subject of much analysis and discussion, it contains mysterious elements hard to understand, tempting the commentator to venture into the dangers of what Matthew Arnold might have called 'metaphysics tinged with emotion'. What does in fact happen? A harsh din breaks out in all the wind. Rebuked by the dignified protest of 'cellos and basses, it is not subdued, but recurs, this time, answering the first notes of the recitative and mollifying its diminished seventh in the direction of B flat (the key we have just left at the end of the *Adagio*) in which key the recitative continues to be diverted to the dominant of D minor where the spectre of the first movement materializes; not the opening, but the recapitulation, for the fifth A–E now rests on a C sharp bass in the remotest depths (double-bassoon and string basses). The scherzo materializes, on the dominant of A minor and is also repeated, in a shorter phrase in F major, ending on E flat, so that it merges into the two bars of the *Adagio*. The recitative is now accompanied and in a remote tonality reminiscent of the episode with the horn solo in the slow movement. Instead of the expected formal cadence in C sharp comes the first phrase of what evidently is a possible new theme in D major. By using the traditional style of recitative Beethoven makes an articulate speech in the strings more eloquent and exact in rhetoric than any words. The strange rhythm of the formal cadence suggests that it would continue if not abruptly silenced by the theme so anxiously awaited. Its 'perfect diapason' is worthy of Milton's lines and only a great poet could find fit words to praise the melody itself and the harmony which grows round it as it is repeated by strings and bassoon and then in the full orchestra. When the last phrase expands in the manner of a transition and the harmony begins to move towards the dominant there is no reason to expect a dramatic crisis for, unprecedented though the beginning has been, Beethoven never repeats himself and we might well expect that, with the resolution of the conflict symbolized by the recitatives and with the perturbed spirits of the earlier movements laid to rest, we were now embarked on a triumphant finale, like that of the Fifth symphony. This is not to be and chaos breaks loose in an harmonic anarchy far more extreme than the din after the *Adagio*. Violent though it is, it yields at once to the human voice. The meaning is unmistakable. (At the words '*nicht diese Töne*' the violins move through a chromatic phrase surely meant to remind us of the coda in the first movement.) Great as is the power of music, embodied in the great theme of joy – we have yet to learn that this is its purport – it cannot alone subdue the destructive and life-denying forces, which melt away at the first sound of the Word. The theme is treated with audacious simplicity in the earlier vocal variations, and the symbolism of the dominant pedal should on no account be missed. The insistent A which had begun worlds ago the tragedy of the first movement now runs happily through the whole course of the 'Joy' theme as the baritone sings it in the lowest part. As remarked earlier, those who set art above life will be offended by these variations and there is nothing useful to be said to them, beyond Schiller's and Beethoven's admonition (see bars 289 *et seq.*). At first the joy is purely human and earthly, but at the lines celebrating the joy, 'Nature provides for all living creatures', a trill-and-turn motive begins *sempre p* in the 'cellos; it grows in power through all the strings until it is predominant in the accompaniment to the chorus. The words '*und der Cherub steht vor Gott*' are declaimed in massive chords, evidently modulating to the dominant. Beneath the unison A the bass falls to F natural; this overwhelming modulation is the realization in the human world of the finale of the plunge into D flat in the *Adagio* (bar 133).

The sublimely grotesque march that emerges from the depths of this modulation is as offensive to some minds as similar things in Shakespeare were to Voltaire. The case is not one for argument; Beethoven's idea of joy is not limited to the sentiments of a liberal-minded assembly and had he felt scruples about offending refined sensibilities he would not have scored this section with the percussion known to the Viennese masters as 'Turkish music', and piccolo and double-bassoon added to the already heavy wind-band. The strange orchestral *fugato* into which the march develops is presumably the musical equivalent of Blake's 'energy is eternal delight'. Not only in key does it foreshadow the great fugue of the quartet, Op. 130. B flat yields to sharper keys and the tone diminishes through a long held F sharp to *pp*. The first three notes of the theme, first in B major then in the minor, are enough to bring in the chorus with a primitively harmonized statement of the theme, accompanied by the 6/8 figure of the orchestral episode. In a kind of triple measure scarcely heard since Palestrina, tenors and basses in unison, with the trombones reserved since the trio for this moment, declaim a new theme, taken up by the whole chorus in four-part writing of astonishing plainness and strength. To make the sentiments of Schiller's text ring true here you must have music as unadorned as the language of the New Testament, and Beethoven does not fail at this supreme test of truthfulness. Structurally it is noteworthy that when the D major chord comes as the dominant of a modally inflected G minor it is followed by a great climax on a chord of E flat. When this chord yields to a mysterious diminished seventh over a pedal A (note the very high double-bass *tremolo* beneath the *pp* wind and strings), the key might be moving almost anywhere, but everything now is centred on D major, so the double fugue, uniting the praise of 'Joy, daughter of Elysium' with the great gesture of 'Let me embrace you, millions', takes its course in the tonic, though with modal features in the harmony. At the climax the dominant pedal comes yet again as a unifying element in the architecture of the whole symphony when the two themes are heard together for the last time in alto and bass beneath the high A of the sopranos, maintained for twelve bars. Whatever Beethoven's religion should be called – and we had better recognize that he was one of the most religious of all great artists – he had no use for attitudes of abject self-abasement adopted in the hope of averting punishment, so it is characteristic that the sudden *pp* which follows the climax of the fugue is to express 'Do you prostrate yourselves then?' – this vital question is perhaps unconsciously bowdlerized into conventional religiosity in the usual English translation – followed by the exhortation 'Seek your

God above the stars, where he surely dwells', i.e. stand up and be men. What can fittingly come after the solemnly radiant passage in which these words are sung to plain block chords accompanied by wood-wind phrases derived from the rising notes of the theme's opening bar?

Beethoven was, it must be confessed, as simple-minded as, say, St Francis, and the last words of this section are '*muss ein lieber Vater wohnen*'. Now if there is a father there must be children, so the solo quartet, in pairs of thirds, sing the beginning of Schiller's Hymn to music of devastating innocence, breaking into a round so simple that the chorus join in as if they had picked it up by hearing the tune. Wagner surely remembered the diminution of the main theme here when he wrote the apprentices' motive in *Die Meistersinger*. This dangerously naïve section gives human expression to the music of the trio, a point made unmistakably by the long tonic and dominant pedals which support the voices. Even at his most '*feuertrunken*' Beethoven never ceases to construct and the horns after holding the dominant for eleven bars in an unbroken *crescendo* break the next three bars into minims, thus anticipating the texture of the coda (see bars 798–800 and 880 *et seq.*). The depth of experience without which the jubilation would be not innocent but merely ignorant becomes explicit at the words '*Alle Menschen werden Brüder*'. The most naïve subject in the work, that which had begun the whole section, is transformed into four bars of *Poco adagio* with a lyrical turn and a brief violin phrase as moving as certain 'ultimate simplicities' in Shakespeare. The brotherhood of man is not something accomplished and present, but depends on the presence of joy. This, of course, must be said by the chorus, but after a brief resumption of the general rejoicing the solo voices dwell on the same thought, moving into B major, the key briefly proposed but not adopted at the end of the orchestral interlude. The *cadenza*-like texture of this wonderful passage may be seen as the outcome of the horn *cadenza* in the *Adagio*. B major turns into B minor and the bass falls a whole tone, a further confirmation of the connexion with bars

529–542. This falling interval is the material of the brief *accelerando* before the last extended climax. Beethoven has extraordinarily few examples of a quickening of tempo, and it would be unthinkable for him to use such a device during the building of a climax, so the *Prestissimo* is reached in eight bars, and maintained to the end, except for the final stroke of genius, when he makes the last choral phrase in four bars of *Maestoso*. He was not a master of words, but to crown the whole work even verbal inspiration visited him with the reversing of Schiller's first line, so that the text ends with '*Tochter aus Elysium, Freude, schöner Götterfunken*'.

4

FRANZ BERWALD

(1796–1868)

Robert Layton

*

On one occasion towards the end of his life, Franz Berwald was asked what he was. He replied not without a trace of irony that he was in fact a glass-blower! Now Sweden is perhaps more famous for its glass than it is for its music and that is hardly surprising when one thinks of the shabby treatment she meted out to her greatest composer. For not only was his music neglected and unplayed during his lifetime, but it was left to gather several layers of dust after his death. In fact the discovery of his music belongs to comparatively recent times and although his stature is now acknowledged in Sweden, international recognition has been slow in following.

His neglect during his lifetime is not difficult to explain: it arose from ingrained habits of mind and outlook, for the idea that peripheral cultures could produce anything other than music of local interest was so novel as to be revolutionary. Who had ever heard of a Russian composer before Glinka or a Pole before Chopin? Of course, these countries and Scandinavia, too, had interesting folk cultures but their isolation from the main centres of musical activity on the continent long retarded any serious development of the art. Such music as there was, largely duplicated musical developments in France, Germany, and Italy. Even Johan Helmich Roman (1694–1758), the 'father' of Swedish music, as he is called, wrote sinfonias that hardly differed from Geminiani and Telemann in accent and manners. They would have remained just as different (or just as similar) had he lived in London or one of the small North German courts. Similarly, in the nineteenth century, the Danish composer, Gade

(1817–90) who wrote no fewer than eight symphonies, modelled his outlook on that of the Leipzig school. Their musical procedures were transplanted, as it were, to the alien situation of a Nordic composer and the result has a certain academic pallor by the side of Berwald.

Berwald reveals many of the influences that one would expect from a composer born five years after Mozart's death and who died one year before Berlioz. But, at the same time, there emerges a personality quite unmistakable in its originality; an innovator in some respects, a conservative in others. His family had a long and flourishing musical tradition behind it and a number of them had held musical appointments at various German princely courts. His father went to Sweden as a young man and settled there. As a boy Berwald was a competent violinist and from the age of sixteen he played in the *Hovkapellet* (the Court Orchestra), leaving it in his early thirties so as to devote himself to composition. Nevertheless he was forced to fall back on his other gifts to make a livelihood; fortunately these included a good head for business and a resourceful practical sense.

Of course, Berwald is not unique in leading this kind of double life as a man of affairs and a composer: indeed, it seems a commonplace among peripheral musical cultures in the nineteenth century. Borodin was a professor of chemistry and other members of 'The Five' were active outside music, while at the end of the century Ives, in America, combined the life of a successful insurance executive with composition.

Berwald's activities, however, were unusually diverse. In the thirties he founded an orthopaedic institute in Berlin where he was then living, and later in life, after failing to secure a musical appointment in Stockholm or Uppsala, he accepted a post in the north of Sweden as manager of a saw-mill and glass-works, an enterprise in which he again showed his flair for handling practical matters. His other interests ranged from publishing to forestry preservation, popular education and rent control!

Although he composed throughout his life, most of the music for which he is best known was written during the

1840s, after he had given up his work in Berlin, sold his institute and returned to Sweden. The four symphonies all date from this period, two being written in 1842, the *Sinfonie Sérieuse* and the *Sinfonie Capricieuse*, and two in 1845, *Sinfonie Singulière* and the Fourth, in E flat. Only the *Sérieuse* was ever played during his lifetime, in 1843: the *Singulière*, which is undoubtedly the finest, had to wait until 1905 for its first performance. Yet the opening bars are quite unlike anything else in the music of its time, or indeed, any other time. The harmonies in themselves are not in advance of Mendelssohn or Schumann but their treatment reveals great originality and imagination. Berwald's classical sympathies are obvious and even though the language of the symphonies does not extend the vocabulary of music, his is a new voice of great freshness and individuality.

80

This is obviously the work of a highly individual mind. For not only is this idea fresh and original in itself, but as the opening of a symphony it is an astonishingly imaginative stroke. Not the least aspect of its individuality is the extraordinary simplicity of means. Who would suspect that it could generate this?

81

Berwald's years as an orchestral player in the *Hovkapellet* naturally familiarized him with the classical repertoire, and the diatonicism of his invention reveals the extent of his debt to the Viennese school. This is not to deny that opera left a strong imprint on some of his orchestral output (the Weber-ish opening of the symphonic poem, *Ernste und heitere Grillen**) but operatic devices, including recitative, which certainly influenced Berlioz and many others besides, made little impact on the symphonies. Berwald's phrase structures

* 'Serious and cheerful fancies.'

are by contrast rudimentary in their simplicity, often being constructed from small segments, equal in length and very often rhythmically identical. This serves to throw added weight on the line itself, as it does, for example, in the effervescent second subject of the finale of the E flat symphony where Berwald's lyrical powers are seen at their finest. This figure which continues Ex. 80 is a case in point.

82

In this respect Berwald cannot always escape the criticism of resorting too readily to sequential repetition, although his sure command of the lyrical graces usually saves the day. Despite this squareness in phrase structure, Berwald is by no means devoid of subtlety in his use of syncopation or carefully contrived shifts of rhythmic emphasis as the following examples from the Promethean finale of the *Singulière* show.

83

Berwald's very first essay in symphonic form dates from the 1820s; only the first movement survives and that is not complete. Nevertheless, there is enough of it to leave no doubt that Berwald modelled his work on the classical symphony. The influence of Beethoven can be seen in the transitional passages where Berwald hammers away at a short, pregnant motive with the insistent tenacity of that master. One characteristic feature of the later Berwald symphonies also emerges in this early A major symphony: the first subject is always heard again before the second group is arrived at and after an attempt has been made to modulate towards the key area of the second group.

Berwald subsequently disowned this early symphony and his first mature work in the form is the G minor, *Sinfonie Sérieuse* (1842). Its title should not be taken too literally. There is, in fact, nothing particularly grave or solemn about it, though it does strike a deeper vein of feeling than, say, the *Capricieuse* or the E flat. It has, too, a greater contrapuntal interest as its noble first subject shows:

84

The second subject of the finale is another example of Berwald's sympathy for the Viennese classics. It is obviously a not-too-distant relation of the second subject of the first movement of the 'Eroica'; yet it even looks forward to Brahms:

85

Berwald remained unaffected by the contemporary movement towards folk-song which gathered increasing momentum during the nineteenth century. He never drew on the rich fund of folk music that is to be found in Sweden, the sole instance in his orchestral music of a direct quotation from folk sources is to be found in the symphonic poem, *Reminiscences from the Norwegian Mountains*, written at about the same time as the *Sérieuse*. Yet at the same time there is almost a suggestion of folk song or at any rate a distinctly rural flavour about this motif. In its orchestral garb it looks forward to Dvořák, who was incidentally scarcely more than a few months old when this was written:

86

The *Sérieuse*, then, was highly personal and forward-looking even though it was deeply committed to the classical tradition. To gather the extent of Berwald's contribution to the symphony and his approach to problems of design generally, it is worth while to take a brief glance at the contemporary symphony. Beethoven and Schubert had brought the classical symphony to great heights; they had increased its dimensions in every way, and invested each of the movements with the feeling of being part of a whole. The last movement in particular emerged from the comparatively light-weight position that it had enjoyed in early Haydn into a movement of much greater significance. The romantic symphony with its programmatic flavouring, its expanding canvas and more rhapsodic content posed new problems of integration. The overriding romantic concern for and interest in detail led to a weakening of structure and a tendency for the symphony to encroach on the realm of the symphonic poem. Berlioz in the quest for unification adopted the principle of an *idée fixe* in such works as the *Fantastique* and *Harold;* even those who followed more conventional lines made efforts to bind independent movements closer to each other.

Berwald's symphonies are roughly contemporaneous with Mendelssohn's 'Scotch' symphony, the first draft of Schumann's D minor symphony as well as the so-called 'Spring' symphony. Yet neither Mendelssohn nor Schumann made so immediately striking a contribution towards the development of an integrated and unified symphonic structure as did Berwald. One of the original ways in which Berwald handled this problem was by enveloping one movement within the folds of another. This is a consistent feature of Berwald's formal thinking. After the scherzo of the *Sérieuse* Berwald immediately recalls the material of the slow movement before plunging into the finale. In the *Singulière* the scherzo is completely integrated into the slow movement, a procedure which has few if any precedents. In the quartet in E flat (1848) he takes this a stage further and encloses the slow movement itself (with the scherzo sandwiched in the middle) in the course of the first movement.

In the *Sinfonie Capricieuse* Berwald resorts to a more common structural device: that of a largely monothematic first movement. This has many parallels at this time including the Schumann D minor symphony (1841) and the same composer's piano concerto. The difference between the material of the first and second groups is very slight; at best, one could call the second group a variant of the first.

87

In the development of material in the symphonies, Berwald occasionally telescopes the first subject when the time comes for the restatement. This happens in the first movement of the *Capricieuse*. This is admittedly common practice in the period: one finds it in Chopin and in a work like the *Symphonie fantastique*. However, he takes even this a stage further in the E flat symphony by introducing a totally new idea in the recapitulation of the finale. No two ideas could be more contrasting than the first subject itself and the theme which replaces it:

88

This stroke in the E flat symphony has few, if any, precedents. In the overall proportions of his design Berwald does not differ greatly from the Leipzig school. The disposition of material in the exposition in all the symphonies is classical though Berwald, as we have observed, tends to delay the second group. In some of the symphonic poems, which conform to the principles of the Mendelssohnian concert overture rather than to the symphonic poem as cultivated at Weimar, he delays the second subject for a considerable time. (In *Wetlauf*, written at the same time as the *Sérieuse* and the *Capricieuse*, it does not appear until bar 222.)

The *Capricieuse* does not survive in its final form: the full score disappeared from the composer's bookshelves shortly after his death and the work as we know it is based on a sketch in short score with orchestral indications, realized by Ernst Ellberg in the early years of the present century. The sketch bears three inscriptions, *Symphonie pathétique* (and this a half century before Tchaikovsky!), *Symphonie singulière* and in larger script and at the head of the page, *Symphonie capricieuse*. The date on the short score also corresponds to a date in Fru Berwald's diary at which time she recorded that her husband had finished a work by this title. How far Berwald subsequently altered the work we have no means of knowing but it would be unreasonable to assume, as does Sten Broman, that there are two separate works called the *Capricieuse*. It is on the strength of this, for Berwald's son maintained that the short score and the finished work were not identical, and the early A major symphony that Broman numbers the symphonies thus: 1, in A major (incomplete); 2, in G minor (*Sérieuse*); 3, in D (*Capricieuse*); 4 (the final *Capricieuse*), 5, in C (*Singulière*); 6, in E flat. On what evidence he bases his assumption that the final version of the *Capricieuse* (so wholly different as to warrant a separate number) comes before No. 5, has never been made clear. Berwald himself disowned the early A major symphony in a letter written in 1829 and subsequently headed the E flat symphony, No. 4. The numbering of the symphonies should therefore read: 1, *Sérieuse*; 2, *Capricieuse*; 3, *Singulière*; 4, in E flat.

Although the question of numbering is not in itself important, it is well worth while recalling the background of the *Capricieuse* for purely musical reasons. It does, in fact, fall slightly below the other three symphonies in the quality of its inspiration. The listener is made conscious from time to time of Berwald's basic weaknesses, his excessive reliance on sequential repetition and the squareness of his phrase structure. In the other three symphonies the vitality of his ideas carries these inherent weaknesses in its stride. The monothematicism of the first movement defeats its own ends by producing a uniformity of contour which throws the symmetrical phrase structure into relief. But this is not to deny that there is some fine music in the work. It has an air of freshness and the finale contains some powerful and individual writing. Although the slow movement superficially seems to belong to the world of Mendelssohn, the fact remains that on closer scrutiny and aural familiarity it speaks with a distinct and individual voice. The chromaticism has a specifically Scandinavian flavour: the passing notes do not produce the cloying flavour that one finds in a good deal of the music of that period, and even occasionally in Mendelssohn:

89

The *Singulière* and the E flat symphony were both written in 1845 and are dated March and April respectively. Of the two the E flat is the slighter. It has a guileless charm that is immediately winning and although it does not plumb great depths, it is more than the excursion into conventional classicism that it seems on first acquaintance. The opening with its stuttering B flats is typical Berwald as, for that matter, is the theme itself, a succession of rising thirds followed by a gentle chromatic descent. The second subject offers another example of his fondness for syncopation: this tune glistens with a delicate wit.

90

The slow movement is based on a theme from an unpublished keyboard work, *En landtlig bröllopsfest* (*A Rustic Wedding*) written the preceding year and its symmetrical phrases reflect something of the well-regulated emotional temperature in which Berwald's inspiration was conceived. Like the slow movements of the other symphonies it is broadly designed on ternary lines. The scherzo has a similarly ordered, symmetrical outlay which betrays Berwald's sympathy for the Viennese classics. This movement has, it seems to me, a Haydnesque lightness of touch and a clarity of texture that mark it off from much music of the time.

In spite of the fact that he lived for some twenty-three years after the completion of the *Singulière* and the Fourth symphony, Berwald never returned to the form. This was undoubtedly due to the negative response with which his scores met. The almost total neglect of his symphonies may have propelled him towards chamber music during the 1850s. Here the question of performance presented no difficulties.

What in sum is the extent of Berwald's contribution to the symphony? If the measure of his importance was the impact he made on succeeding composers, Berwald would be a negligible figure, for his work lay undiscovered in a country remote from the mainstream of music and had no opportunity of exercising any real influence. If, too, the yardstick by which he should be measured is the extent to which he contributed to the development of the language of music itself, his achievement is a comparatively marginal one. Berwald belongs to the category which is less concerned with the enrichment of musical language by an expansion of its vocabulary than with the individual use of already existing expressive means. And in this lies his fascination. His formal innovations are bold and leave us in no doubt that he brings a fresh approach to the symphonic problem. His apparent conservatism in matters of tonality* and in his harmonic vocabulary do not prevent us from being surprised by many personal touches that remind one of Weber or look forward to Dvořák or even Reger. Berwald remains obstinately original.

Every composer creates his own world. Berwald's is not a world in which the heroic, the epic or the profound play a very great part. They lie well outside his horizon. Not for him the white-heat of, say, Berlioz's *Harold* or *Romeo and Juliet* symphonies: the limits of his emotional orbit are more carefully circumscribed and in some of the slow movements of the symphonies (particularly the *Singulière* and the E flat) though we find warmth, compassion and humanity there is none of the feverish intensity of Romantic art. Yet if Berwald is not a composer for whom exaggerated claims should be made, we must acknowledge the fact that within well-defined limits he moves with complete and compelling mastery.

* His early works evinced a remarkably audacious feeling for modulation but the mature symphonies are far less bold in this respect.

5

FRANZ SCHUBERT

(1797–1828)

Harold Truscott

*

1 in D major	1813	
2 in B flat major	1814–15	
3 in D major	1815	
4 in C minor	1816	The 'Tragic'
5 in B flat major	1816	
6 in C major	1817–18	
7 in E major	1821	Incomplete
8 in B minor	1822	The 'Unfinished'
9 in C major	1828	Often called 'The Great C major' to distinguish it from No. 6.

SCHUBERT's earliest orchestral compositions show, as clearly as his later works, that there is no such thing as typical Schubertian orchestration. His orchestral style changes, even in music written when he was fourteen and fifteen, from work to work. In each, it is complete, unified and unique. The nucleus of his orchestral styles is the brass, horns and trumpets to begin with, later trombones, which he uses with unprecedented freedom and variety of approach. In each work the style is communicated to the brass and from there to the rest of the orchestra. In this he seems to have been guided by an innate stylistic sense, performances only verifying what his imagination had already taught him. Above all, his orchestration is always conceived as part of the whole creation, not as a dress added after the body is complete. At the same time, his orchestration is only the outward and visible sign of a unity inherent in the matter, or content, of his music. Without this it would be an affectation: with this it is unselfconscious and natural.

The content of Schubert's work often inspires a dangerous simplification. There are many levels to his thought, and as a rule only one is considered. He is indeed a tuneful composer, but it is not difficult to show that, in a large part of his early work, the tunes or themes were what interested him least. They were a matter of convenience, since he felt that there must be in music some kind of linked connexion between one note and another. If one examines some of his early instrumental music one finds a peculiarity about the majority of his themes: they are not his. He borrowed almost as heavily as Handel. One is left with the impression that any theme will do and, for his purpose, this is near the truth. Individual melodic invention did not come till later. In his early music, the source of the themes is irrelevant. The movements Schubert builds on them are purely Schubertian.

Earlier composers gave much more prominence to their themes than we find in any early instrumental work of Schubert. Can one imagine the first movement of Mozart's last E flat symphony without the actual thematic material which informs its structural processes, as well as contributing largely to the beauty of the music? Can one imagine the first movement of Beethoven's 'Pastoral' symphony without the particular theme with which it opens? A large part of the progress and atmosphere of the movement depends precisely on this theme and on no other imaginable one. But listen to the first movement of Schubert's Second symphony and gauge how much it relies on its main theme. If any music can move in defiance of melodic and thematic control, this is it. So little was Schubert concerned with themes for their own sake that he made use regularly of the same germ, the main theme from Beethoven's *Prometheus* overture. Theme after theme in these early works is derived from it, and is taken through the same local harmonic process, and yet no music could be more different from Beethoven's. But here we are up against another stumbling-block.

Schubert is often spoken of as the last of the classical composers, with a strong leaning towards the romantic.

The truth is that his music is as nearly the opposite of classical as is possible, yet we need a new word, for 'romantic' will not do either. Quite naturally, Schubert began by using the outward forms of the music on which he was brought up. What has scarcely been noticed (except as ineptitude) is that Schubert never uses those outward forms in the way that all classical composers use them. Concentration on Schubert's tunefulness and disappointed expectation of a general conformity to classical practice have led to the conclusion, now generally accepted, that Schubert was not bothered with form. Had attempts been made to discover what positive form Schubert's music does achieve it would have been found to be entirely unclassical. The truth, when all is taken into account, is that Schubert was interested principally in form, in the sense of articulate personal expression of complete thought. But it is not classical form, and the fact that it masquerades for so long in a classical dress has proved to be highly misleading.

It is useless to assume that such a movement as the first in the D major string quartet (D. 94), composed when Schubert was sixteen, is simply a ham-handed attempt to handle classical sonata style, when the sheer sound shows that it is, as music, completely convincing. It is a positive form and it is quite unclassical. Evidence piles up from the very beginning; from, for instance, the earliest orchestral work we possess, the overture *Der Teufel als Hydraulicus*, composed when he was about fourteen. Here there is scarcely any material to be concerned with. Here are its main themes:

91

Neither of them would win any competition for tuneful invention. The piece itself is almost static, being practically confined to its tonic chord of D major, with the immediate relations. (There is one deviation in the middle to Schubert's favourite flat submediant harmony, but this is short-lived.) Schubert's individuality here has two bases, which become gradually more important in his music. One is his variety of phrasing – the reverse of lyrical. The other is precisely the rooting to one harmonic spot. This is not the result of helplessness, but is intended to serve some purpose when the time is ripe. Further, the young Schubert displays impressive control.

This overture is one small symptom of what is to come. In all the early work we find classical processes used apparently enough for classical terms to be convenient in description, and yet in a manner almost directly opposite to that of the classical composers, to produce completely opposite results. For instance, the First symphony, composed in 1813, begins with an *Adagio* introduction followed by an *Allegro*. The surprise comes when Schubert precedes his recapitulation with the music of the introduction. If his directions are followed, there should be no slackening of the prevailing *Allegro* tempo to accommodate this music, and yet the suggestion of the *Adagio* tempo shines through. It is not exact, but it sounds so at the distance separating the two passages. At the return of the *Allegro* a few bars later the beats move at the same pace in bars of the same size, yet the difference in tempo is obvious to the ear.

In this way Schubert provides the introduction with a connexion with the main movement other than that of bald contrast. This particular feature, however, is a result of a much larger process forming, admittedly with only partial success, the whole movement. The object is clear enough. The music contains a duality of pace – two speeds running concurrently. This device is found in much later nineteenth century music, in Wagner, for instance, but not in any music prior to Schubert, that of Beethoven included. Schubert felt this instinctively at first: increasing technical maturity made him consciously aware of it. Before he achieved this consciousness he used the classical *Allegro* as a matter of course. His later music proves that his object was to slow down the classical *Allegro* pace (for a reason which will become apparent), and so almost automatically that pace collides with another, contradictory, one. (Schubert shows in certain works, the Fifth symphony, for instance, that he is quite capable of handling classical momentum with classical mastery; but this is not his main trend.)

Several points of Schubert's early style now fit into place, particularly the stretches of non-thematic writing. The length of these is without precedent in the music of classical composers. Paragraphs of 20, 25, 30 or more bars of pure non-thematic writing abound in the first movement of this symphony alone, and it is not an abnormally long movement. Even where there is a thematic thread, it is reduced during long stretches to the reiteration of one short figure, usually his favourite dotted rhythm. Such procedures are not compatible either with a classical sense of sonata or with a naturally lyrical mode of expression. Also, since the movement itself is for most of its length immensely convincing, one cannot put this process down to ineptitude.

This device of non-thematic writing is associated more and more with maintaining harmonies over long stretches – far longer than is compatible with classical momentum – and thus is connected with the gradual slowing down of pace. We find it in the two overtures in D (D. 26, of 1812, and D. 556 of 1817) and the two 'in the Italian style', also of 1817, as well as in the first and last movements of the first four and the Sixth symphonies. Progressively more able use of it is to be found in the corresponding movements of the contemporary string quartets. The music itself assures us of Schubert's definite plan – instinctive at first but later consciously controlled – if we will examine it as a continuous growth.

So much depends on the degree of awareness with which each of us listens to any music, particularly when it can be easily taken, like Schubert's, to be in a style with which we are already familiar. The Fourth symphony is an excellent example. After its impressive slow introduction its *Allegro* moves with a recognizably classical speed of movement – or does it? Compare the opening of this *Allegro* (Ex. 92(*a*)) with the second subject in the first movement of Beethoven's C minor string quartet, Op. 18, No. 4 (Ex. 92*b*):

92

and with the speed of that movement as a whole. The thematic connexion is so obvious as to need no comment. The Beethoven moves at a definite quick speed occasioned by the use of each crotchet beat of the bar as a fundamental impetus. This is what Schubert appears to be doing, but it is mainly appearance. His *Allegro* is there, but it is mixed with another element, absent from the Beethoven. The rhythmic stress of the Schubert also depends on the length of time he maintains one harmony, and there is a doubt set up as to whether the speed is not a slow one, after all; and this slow element gradually assumes the stronger pull. The matter is put beyond any doubt by the beginning of Schubert's development:

93

where we are presented in quick succession with three different speeds, in an enlarging scale, from a classically quick one of four quick beats to a bar, then slower with two minim beats, to the slowest and most predominant in this movement, of one large beat to each bar, sometimes stretched to two bars. These speeds co-exist. This is a new resource in the writing of symphonic music, which has the effect of making the music as different as possible from classical music in its means of obtaining its object, although there is a surface similarity. In fact, this similarity makes the difference all the more potent, which may be why Schubert retained classical dress for so long.

But this process is not isolated; it is part of a pattern. The basis of Schubert's thought on a large scale is tonal and harmonic. He often extends one harmony over a number of bars, so that a passing harmony can actually carry whole phrases. He increasingly used the Neapolitan harmonies* in this way, on both tonic and dominant. (Finally, as we shall see in the great C major symphony, the two are brought together in one continuous statement.) The more one hears of Schubert's planned harmonic and tonal operations the more apparent it becomes that their full working depends on the gradual slowing down of the traditional classical *Allegro* pace, without lessening the quickness of dramatic thought. Indeed, the thought moves so fast that the slower pace is essential for its comprehension. It is a paradox, but Schubert found it necessary to go slower in order to think faster.

Often this use of the Neapolitan harmonies is concerned in the establishment of a new key, as in the D major overture in the Italian style, where the first subject moves to the dominant, and the A is selected as mediant of an F major chord, with which the second subject begins. This is Schubert's transition harmony, which eventually moves inevitably to the real tonic, A, sensed for many bars before it appears. It was with such avoidance of traditional methods of key-change that Schubert began working towards that personal type of tonal drama which we find fulfilled in the later piano, chamber and orchestral music. It has earned him among people who use their eyes more than their ears the erroneous reputation for dealing in abrupt transitions. He begins by this means to put the strain of his structure on the second group – because that is where his tonal drama is going to work itself out – rather than the first, so that his second groups become larger and his first groups consistently shorter. A further cause of misunderstanding is the teaching that the classical sonata is essentially a style dependent on melodic contrast. Even the earliest examples do not support this view. Schubert, in beginning a second group of themes at one point and achieving

* The Neapolitan Sixth is the chord a semitone higher than the tonic, or key-note, leaning on that tonic, usually in first inversion. It can also be found on the chord a semitone higher than the dominant–the fifth note of the scale.

his real change of key at another, was only taking the real classical structure to its logical end, although his methods and results are very different from those of any classical composer.

The extraordinary shape of the Second symphony's first movement is one of the first results of this structural process. After a stately slow introduction the main theme

94 

(perhaps the chief of all Schubert's variants of Beethoven's *Prometheus* overture theme) moves in beautifully uneven phrasing to a transition theme, formed from the quavers of Ex. 94, starting from C minor harmony and moving, still in uneven phrasing, to E flat major and a lovely tune to begin the second group, with Ex. 94 quavers accompanying. These quavers, in a variety of shapes, prove to be the only theme of basic importance in the movement. The tune is left, the quavers form a whirling transition to F major and Ex. 94. From the beginning of the movement, except for the E flat tune, the quavers have been the only thematic material, which supports my statement that Schubert is little concerned with themes. Two sets of cadence chords come at a fairly wide interval, and again we get the impression of two speeds in operation. The exposition, however curious its shape, ends on the dominant of F.

The development, shifted up a semitone to D flat (a Neapolitan suspension) makes sequences on the second set of cadence chords accompanied by the quavers, and arrives at the recapitulation – in E flat major. Schubert does not restate exactly, as his key would allow him to do, and both the original E flat tune and the F major statement of Ex. 94 now come in B flat major. So that his purpose in this movement, in which the classical divisions are nominal, is tonal and harmonic, not thematic at all. This fusing of two keys into one is another gradually developing feature of his style, never appearing twice with the same meaning or result.

The structure of this movement is quite as epoch-making as that of many acknowledged milestones in the history of musical form, but it is achieved with such consummate art, with such a light touch and persistently beautiful sound that it passes us by. We take it at its face value, as we often take people we meet. And yet the structure is of vital importance. The more we listen the more we shall find plan, structure, form at the bottom of whatever Schubert does. The structure requires the intensive listening which he designed it for, and which it seldom gets.

The tonal structure of the last movement is equally remarkable and provides a balance to that of the first. But it is important in another way. It moves, like the first movement, on one fragment of its main theme, a rhythm of a crotchet and two quavers, which is rarely absent throughout the movement. And this provides a heavily accented four-bar rhythm which reappears continually in Schubert's music, culminating in the second subject in the finale of the late C major symphony. It has the frightening sound of sheer physical repetition, which Schubert knew so well how to produce. The main theme is a pretty tune, but we must beware of Schubert when he begins a movement thus. He is preparing a major onslaught. The plunge, without a break, into the development through a whole series of remote harmonies yet entirely in this rhythm, to be followed by the forming of simple sequences leading gradually to the recapitulation, is astonishingly impressive.

I have not referred specifically to the slow movements and minuets or scherzos of these early symphonies, because their beauty of form and treatment is recognized. Here Schubert's lyrical vein has full sway, and he is least challenging to tradition. Their phrasing well repays examination, however; it is not by any means that of a cut and dried lyricist.

Up to his Sixth symphony, with the partial exception of the Fifth, Schubert's symphonic writing presents a consistent development, with certain structural processes assuming more and more importance, as I have outlined. But throughout these works certain harmonic and thematic habits are evolving, which burst out at full and final strength in the C major symphony of 1828. The tendency to build on the German sixth, arriving by this means at a triumphal climax, the insistence on rhythm rather than melody as the main material, the use of a theme building up the common chord in trochaic rhythm (as at the end of the D major Italian overture), harmonies descending in thirds on a resolving seventh, all these contribute to that final statement and all are developed in the earlier symphonies. The finale to the Fourth symphony stands out, superb in spite of certain weaknesses, once again characterized by this overwhelming rhythmic insistence. It is almost a preview of both outer movements of the great C major symphony.

The E major symphony sketch of 1821, a much weaker work than anything earlier (except for a magnificent scherzo) is notable mainly for experimenting in both outer movements with a tonal process which comes to full fruition in the great C major. In May 1818, however, he had sketched out in piano score a symphony in D that promised, if completed, to have been his finest to date. An introduction and part of a first movement, three different slow movements, a scherzo (unfinished) and trio, and three finales each with a number of variants – in this mass of hastily written music lies some of the finest he ever conceived. A B minor *Andante* in 3/8 especially is unmatched in his work or anyone else's. It is the epitome of loneliness, expressed unsentimentally with a bareness of harmony one hopes he would have left untouched:

95

Was there ever a more heart-breakingly beautiful passage than this? Such music demands new terms for both classical and romantic. It is outside either. Its opening led later to a famous beginning in the same key:

96

While Schubert's symphonic style was maturing he was devoting much effort to working out his personal relationship with the piano. By 1819 he had discovered his own way of using this instrument, and he left it for a time. In 1819 also he wrote what is perhaps his finest symphonic movement before the B minor symphony, an overture in E minor. Here, as in the piano music of this year, technical and spiritual maturity meet. This is a movement unprecedented in his earlier orchestral music for its conciseness and its application throughout of the principles which had been his main concern. Here we have the dual tempi, underlined by a much surer tonal grasp. His method is now fully-fledged. He approaches a new key by side-stepping from the home dominant, taking up one note of this chord (B major), turning it into his transition harmony (G major) with the beginning of the second subject. At the same time he reduces his main theme and second subject to one pregnant rhythm:

97 

He moves gradually by means of Ex. 97(*c*) through the most thrilling passage of actual transition he had yet written to the fully established key of B major. It is doubtful if any such passage in his later work is finer in its controlled proportions. The whole overture, as a result of tonality, harmony and the double tempo, is in suspension, depending on perpetually leaning harmonies (nearly always Neapolitan) resolving on to chords which change before they can be asserted, so that only at one or two spaced out points is there any suggestion of a momentary finality. The proportions of this work are almost exactly those of the first movement of the late C major symphony.

After this overture there was a gap, except for the E major sketch, until 1822. During this period he concentrated on vocal music. In 1822, however, he erupted both through the medium of the piano, with the 'Wanderer' Fantasy, and orchestrally with the B minor symphony. In most obvious ways this stands apart from all the previous symphonies and the one that followed. It is to some extent foreshadowed, however, by the 1818 sketches: its opening, for instance, clearly derives from the beginning of the B minor *Andante* (Ex. 96):

98 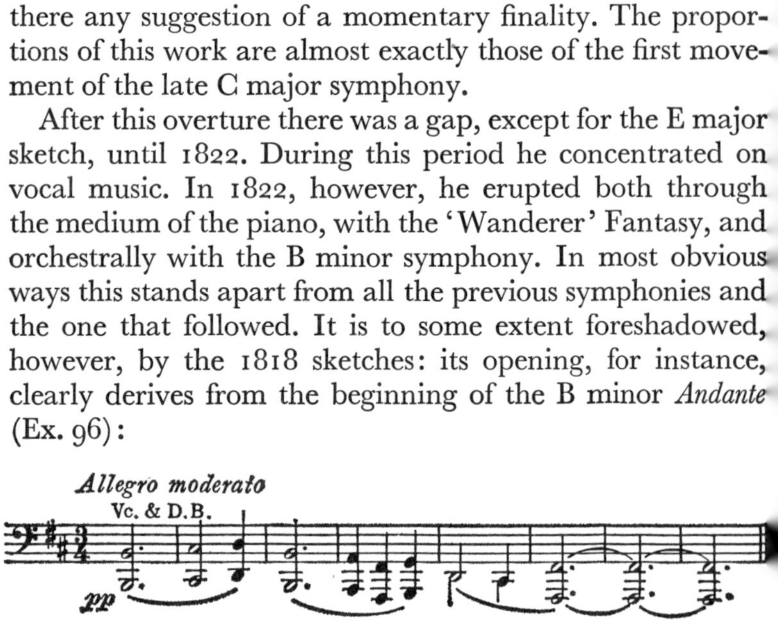

All the trends towards slowing down of pace in the earlier music lead towards this symphony. This is not the most obvious thing about its first movement, but it is one of the deepest and most lasting, and it is certain that without it this movement could not exist. Here the slowing down is complete. There is no more mixture of tempi. We have arrived at the goal, a tempo much slower than any conceivable classical *Allegro*, designed to accommodate thought so dramatically quick that any faster tempo would make it impossible to comprehend.

Few people have escaped the tragically pungent spell of this music. Its themes, completely individual now as those of the earlier works so often were not, are known even where the work itself is unknown. In this country this is the sign of a true classic. Schubert's symphony *is* a classic, although it is the reverse of classical. Here are the important features which contribute to the overall result. First, consider Schubert's beginning. In his earlier music, where his themes are comparatively unimportant, his developments are almost always tonal. In later works, though development of keys and harmonies remains the most important aspect of his music, his themes tend to be individual and important in their own right. Here, in this symphony, we have a development which is equally tonal (it simply drops in steps from subdominant to tonic) and thematic, and an unusual exposition to prepare the way. A first theme, Ex. 98, is followed by a normal sonata exposition. Ex. 98 proves to be the thematic back-bone of the development, and appears elsewhere only in the coda. The exposition proper contains the famous abrupt transition (the only one of its kind in his work) which has its explanation in the recapitulation. There is also the beautiful second subject, in G major, with its accompaniment used tellingly in the development:

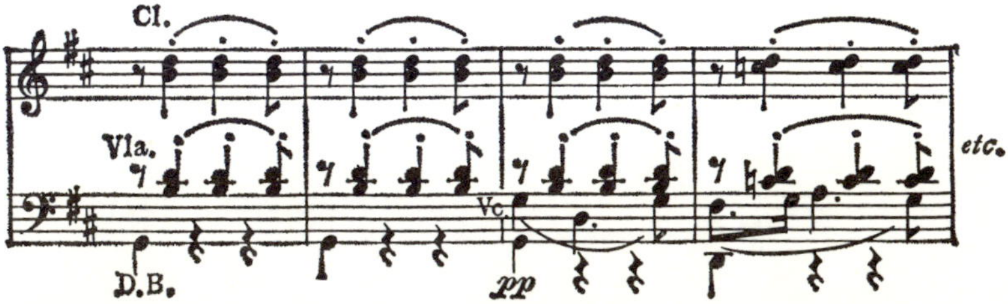

99

So the first subject, apart from the normal first group, bears the main weight of the tragedy, and the exposition sets in motion the tragic catastrophe itself. The enactment of this catastrophe is the answer to the abrupt transition. In the recapitulation Schubert's second subject sings happily in D major, until the silent bar. In the exposition it had remained in G major; here, it is extended to encompass a real transition from D major to B minor. This *is* the catastrophe. This process is the opposite of his more customary one of fusing two keys into one in the recapitulation, and it depends on that 'abrupt' transition in the exposition.

The second and only other finished movement does, in fact, complete the symphony as nothing else could, and I believe that this is the real reason why Schubert did not go on with his partly sketched scherzo. His superb sense of balance leads him here, as elsewhere, to make the movement lyrically simple where the first is dramatically complex. It has three themes, forming one idea, although Ex. 100(*c*) is developed from Ex. 100(*a*):

100

The *tutti* at bar 32 is a cunning double development of Ex. 100(*a*). The episode (*on*, not *in*, C sharp minor) is reached by a characteristic harmonic drop of a third, purely lyrical yet magical. It is evolved from Ex. 100(*a*):

101

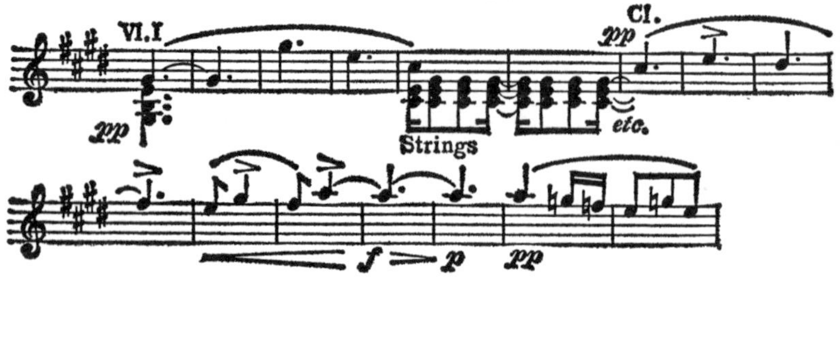

For all its climaxes the movement never leaves its lyrical confines, but it radiates the spiritual strength necessary to sustain the tragedy of its companion.

From this symphony were derived many of Bruckner's most characteristic features and, at the opposite pole, some of Brahms's statuesque but dramatic *Allegros*: the first movement of his Second symphony, for instance.

The great C major symphony is a summing-up of Schubert's instrumental thinking from 1811 onwards. It is no wonder that this symphony flowed out as it did, since practically all the instrumental music he had written was in some sort a sketch for it. Its known features have been often enough described. Rather it is necessary to draw attention, in the first movement, to what happens in the vast second group of themes. Most conductors retard the pace suddenly as this is approached, a procedure which shows unawareness of the continuity of Schubert's thought. The second group begins on the transition harmony of E minor, which disappears almost at once, and approaches the second *key* through the trombone passage, utilizing the second bar of the introduction to do so:

102

This passage is not appreciated if it is highlighted, but only if it is played for what it is: one of the most wonderful tonal transitions ever conceived, arriving at G major finally and fully only fourteen bars from the end of the exposition. The whole second group is one complete tonal transition from C major to G major. This passage is the central fact; from this everything else in the symphony radiates.

And this is caught up immediately in the final linking of

the two Neapolitan chords (see p. 195). The development presents this harmonic and tonal outline:

103

Unfortunately the limitations of musical terminology are such that 'A flat' has to serve for three quite different phenomena, from which Schubert extracts deep meaning in this passage. There is A flat, the flat supertonic of G major (here is the reason for G major being established only at the end of the exposition), A flat, a real and fully established key, and also A flat, the flat submediant of C major. These are linked by falling and rising thirds, producing gradually changing harmony from one point to the next. All is supported thematically by the fact that the first half of the development, coming *from* the exposition, is based on the beginning of the 'second subject', while the second half (after the fully established A flat), going *to* the recapitulation, takes up a fragment from the introduction. These things are not dull technicalities; awareness of them renders the whole magnificently alive. With the recapitulation Schubert extends the harmonic exploration of C major to bring the second group to a complete restatement in one key, C minor. This passes to the major, bringing to his original transition passage, poised around the trombone, a quite new perspective.

Schumann's famous reference to 'heavenly length' (of the *Andante*) shows his perception, unconscious perhaps, of Schubert's perfect form, of all that radiates from the controlling centre of tonality. The repetitions of the *Andante* are the inevitable rhythm of nature controlled by art, the tonality lyrical as always in this type of movement. As in most 'lyrical' Schubert, its phrasing is worth examination. The introductory theme in the bass, a seven-bar phrase:

104

contains the whole main group of themes in embryo, while the oboe tune:

105

is a nine-bar phrase with odd internal subdivisions of two, three, two and two bars. Again and again what begins as a normal four bar phrase becomes six bars, and many times an odd bar stands quite naturally on its own. Once again we have a double speed working, so that the movement can never be either fast or slow, no matter how maltreated it may be in performance. Here, too, more immediately than anywhere else in the symphony, we may note the imaginative treatment of the trombones, which again determine the general orchestral treatment throughout the work.

The scherzo is different from many earlier ones only by its

size, which is commensurate with the work as a whole. It i a full-scale sonata movement, with once again a doubl speed in operation. We may note the echo of the first move ment provided by the shift up a semitone at the beginning o the development. Here, too, the four repeated notes of th finale's second subject make their second appearance in thi work:

106 

We have already heard them in the *Andante*, at the end o the oboe tune (bars 8 and 9 of Ex. 105) and at the beginnin of the second group:

107 

The double speed of the trio has been proved unconsciousl by most of Schubert's commentators, some of whom call i fast, others slow. The truth is that it is both at once, or som speed which is neither. It, too, is a full-scale sonata design and characteristically makes frequent use of the whol phrase supported by passing harmony.

The tonal centre of the first movement operates in its ow way in the apocalyptic vision of the last. The finale simpli- fies the complex tonal situation set up in the first move- ment. There is a straightforward move to the dominant of C the start of the second group on this harmony and a long gradual transition-passage to a real G major, so that the G we finish with is not the same as that we began with. The development sidesteps to a local E flat major and move slowly upward in steps of a fourth to C sharp minor, thence by thirds to the dominant of C. The E flat the development begins with is important, for, after a long preparation for C major on its dominant, the music sidesteps on to E flat again and the recapitulation begins from there, with something of the same effect as in the approach to the trombone theme in the first movement's exposition:

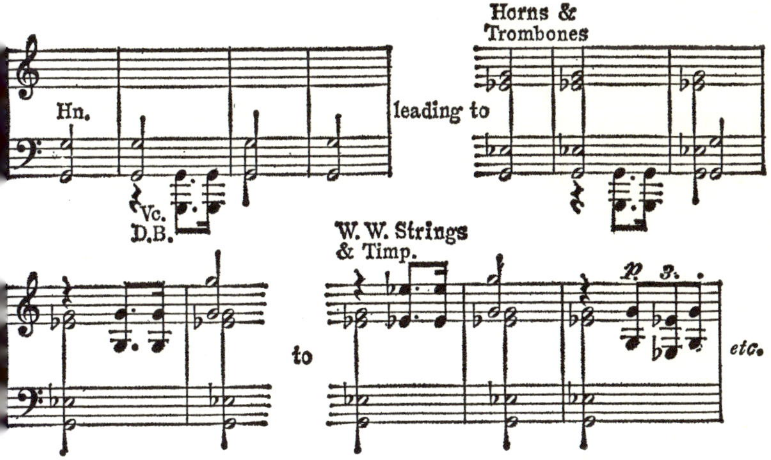

108

This takes us, by means of Schubert's favourite German sixth, to G minor harmony, and the music moves gradually (everything in this movement of tremendous speed happens gradually) to F major, which proves to be a chord leaning on E major harmony. The E major is, and sounds like, mediant of C major:

109

and the whole group follows in this main key of the work. As at the end of the exposition, the music moves down in three steps (not quite the same steps) to the coda, which builds up, from A, the final courageous triumph.

Our final conclusion must be, then, that the spirit of Schubert's symphonic writing is utterly different from that of his classical predecessors. The paths mainly followed by German and Austrian symphonic and chamber composers in the later nineteenth century, different as the two lines are, can be traced back to him. This trend had begun, under a bewilderingly simple surface, in his earliest chamber and orchestral works. Behind the apparently classical language, behind the spasmodic echoes of Mozart and Beethoven, Schubert was slowly but surely encompassing one of the most far-reaching upheavals in the history of European music – as far-reaching as Beethoven's, although quite different. But whereas Beethoven's (although imperfectly understood) is recognized, the fact of Schubert's upheaval is still scarcely suspected.

6

HECTOR BERLIOZ

(1803 – 69)

David Cairns

*

In spite of an apparent formlessness, there is an inherent symmetrical order corresponding to the great dimensions of the work, and this besides the inner connexions of thought. (Schumann)

Formally speaking it is among the finest of nineteenth century symphonies. (Constant Lambert)

Far from being romantic rhapsodising held together by an outmoded literary commentary . . . one of the most tautly disciplined works in the early nineteenth century. (Wilfrid Mellers)

A work of deeply conscious thematic planning . . . . The deeper we delve into his compositional fabric, the more we come to realise that such a view [the 'literary' view] is entirely without justification . . . besides having enriched the orchestral palette with new enchantments and accents, his actual strength as a musician lies in the pronounced sense of thematic consistency, which as a pioneer he managed to transfer into an entirely different idiom [i.e. from Beethoven's] and, in fact, into entirely different patterns. (Rudolf Reti)

It will be obvious that it is the 'Fantastic' symphony that these writers are discussing; but this view has not always predominated in the 130-odd years since Berlioz composed the work. The title 'fantastic', the detailed literary programme which Berlioz at first attached to the work (and which commentators at a loss for an idea have taken up with disapproving relish), and the glamorous, picturesque and grotesque elements in the score have often blinded people to its true character and stature. Certain academics, looking everywhere rather than at the music itself, have discovered that some of the material of the symphony comes from earlier works, and that for this reason it cannot be a unity. Berlioz himself admits that the sad melody for muted violins at the very beginning of the work is the tune of a song which he wrote as a boy (to the words '*Je vais donc quitter pour jamais mon doux pays*'). The first part of the *idée fixe* (the main theme of the work) is lifted from a student cantata, *Herminie*, with which Berlioz unsuccessfully competed for the Prix de Rome two years earlier. The March to the Scaffold was borrowed lock stock and barrel (with the *idée fixe* simply stuck on at the end) from the unfinished opera *Les Francs Juges*. Ergo, the 'Fantastic' symphony is not really a work but an anthology of existing music connected by the current of Berlioz's intensely imagined autobiographical self-dramatizations.*

The view has a superficial plausibility, and it appeals to those who, as Schumann said, are genuinely puzzled by Berlioz's style, particularly his melodic style; but it will not stand up. We must remember that Berlioz in 1830 was a Frenchman writing for Frenchmen. Unlike many great musicians he was an educated man with a highly articulate mind. He had a flair and a taste for describing and analysing music in words. And he exercised this talent to the full not only because he had to in order to earn his living as a journalist (since music did not pay) but because literary expression was a natural and inevitable weapon for the Romantic of 1830. Artistic manifestos were both fashionable and necessary. They were propaganda for the cause, shots in a war in which every victory must be won and registered in public. Men like Victor Hugo, de Vigny, Delacroix, whatever their differences, were conscious of being part of a united movement of liberation; they strove, in an atmosphere of the barricades, to free art from the life-denying rigidity of academicism and dedicated triviality and to give it expression and the beauty of unvarnished

* E.g., in Collins's *Music Encyclopedia*, ed. Westrup and Harrison: 'largely patched together from previous works.' The same kind of thing is sometimes said of Berlioz's second symphony, *Harold in Italy*: a convenient title on which to string together bits and pieces, some of which had already done duty in the discarded overture *Rob Roy*.

truth. It was a movement that needed, and wanted, to explain itself publicly; and it was to this fraternity of writers and painters, not to the Parisian musicians of the day, that Berlioz felt himself to belong. There was also the fact that he was introducing music of startling novelty to a public which had only lately heard Beethoven's symphonies for the first time and was still shocked and divided in its mind by the experience. Besides, as Schumann remarked, 'To the French, music by itself means nothing.' Even without his much advertised passion for the Irish actress Harriet Smithson who had appeared in Paris three years before in Shakespeare (the greatest of all the gods of the French Romantics), Berlioz would have felt obliged to publish a written 'explanation' of his revolutionary music.

Everyone knows how Miss Smithson was in the audience at the first combined performance of the Fantastic and its sequel, the monodrama for speaker, soloists, chorus and orchestra, *Lélio*, and how, according to Berlioz, almost overcome with emotion she murmured to herself: 'Good God! Juliet – Ophelia – I can doubt no more. It is of me he speaks, he loves me still!' But it is the work itself that has to be judged, not the personal and no doubt exhibitionistic uses to which Berlioz put it at one stage of his life. It is less well known that he later removed the literary commentary, as so much scaffolding that had served its purpose and could now be taken down (directing that it should only be used when *Lélio* was also being performed); he did so in the belief that 'the symphony can provide its own musical interest independently of any dramatic intention'. In this he showed sound judgement. While *Lélio*, however instructive and entertaining we may find it, remains as a work of art a prisoner of its time and fashion, the 'Fantastic' is an ageless masterpiece. As the musicologist Rudolf Reti has said, 'the music stands its ground, while the programme evaporates into void'.

If we pay attention to the work and not to what the young Berlioz may have said about it, we come to realize this. There was nothing new in a symphony inspired by an idea. Indeed, what struck Berlioz above all about Beethoven's symphonies was that 'a poetic idea is everywhere manifest', but that at the same time 'music is wholly in command.' The poetic idea of the Fantastic is not Berlioz's love for this or that woman but the 'state of soul' – the intimations of passion as yet undeveloped, young faculties feverishly active but exercised only on themselves, love in search of an object, idealism in an indifferent universe, the aching sense of isolation – which he found echoed from his own experience by such writers as Chateaubriand and Saint-Pierre,* and which seized on an imaginary beloved on whom to project obsessively its vague longings. The 'Fantastic' symphony has survived not because it tells a story but because it expresses, in dramatized but fundamentally symphonic form, the eternal pains and ardours of youth. In the actual process of composition, autobiography was fully absorbed into art. Personal feelings – and in the finale, hallucinations – have been objectified into classical statements.

Another source of misunderstanding is that though Berlioz was a conscious disciple of Beethoven and was profoundly influenced by Beethoven's thematic mastery and consistency, his basic materials and his formal procedures were different; and an attempt to analyse the symphony in terms of sonata form will lead only to frustration. Berlioz in the 'Fantastic' symphony was speaking a new language: not only a new language of orchestral sound, with double basses in four parts, timpani playing chords, a movement begun by a solitary wind instrument unaccompanied (the cor anglais in the *Adagio*,) and the percussion raised almost to a fourth estate of the orchestra – but also a new language of feeling, of extreme nervous sensitivity and poetic awareness, which Wagner recognized, when he heard *Romeo and Juliet* in 1839, as a hitherto unknown world of music, the outward and

* It is curious, in view of the crucial importance of the interval F–E in the Fantastic Symphony, that Berlioz's copy of Saint-Pierre's story *Paul et Virginie*, which he read in adolescence, was annotated by him on the title page with a few bars of music, beginning:

visible sign of which was the unheard of fastidiousness with which nuances of expression were marked in the score. It was not surprising that when Berlioz, with his deeply classical and even conservative cast of mind, sought for the right form for this language ('that form without which music does not exist', as he said to the Princess Wittgenstein) he should arrive at something novel – something which, for all his adulation of Beethoven, embodied Beethovenian principles of thematic consistency in 'entirely different patterns' (Reti). He was also a disciple of Gluck. And instinct told him, even as a young man, how his talents should be used.

We will come to these patterns in a moment. Consider, first of all, the remarkable thematic homogeneity of the symphony. Ultimately it is not this that makes the Fantastic a great work, but it will show how carefully Berlioz has embodied his drama in purely musical terms, and has created this unity organically from the bottom, not at a later stage of composition. The main theme of the work, the *idée fixe*, announced after a slow introduction, extends to forty bars.

110

The features of this theme are: (1) Its length. (2) Its strikingly irregular phrasing: the first 8-bar statement, tonic C to dominant, is followed by an answer, dominant G to tonic, seven bars long. In any reasonably lively performance we feel this irregularity very strongly, and we are likely to be at first disconcerted by it, before coming later to delight in it and to recognize it as a source of strength. Schumann said, 'Modern music has certainly never produced a work in which even bars alternate more freely with uneven', and the force of this fact has not been altogether lost since then. (3) Its monodic character and virtual independence from accompaniment, emphasized rather than diminished by the detached staccato figure on the lower strings which enters at the seventh bar. (4) Its shape, which constantly aspires upwards, only to fall back to the same pair of notes, F-E. This destiny is implied in the opening bars of the theme: the first challenging upsurge of energy (to which the long tied G and the dynamic syncopation of the succeeding four notes give tremendous impulse) is checked at the *sforzato* F and falls back – the musical expression of Chateaubriand's 'faculties exercised only on themselves . . . living with a full heart in an empty world'. Always, despite its struggles to escape from these frustrations, the melody is forced back to the same point.

This kind of melodic sigh of a rising and falling semitone is a characteristic of Berlioz's style and occurs constantly in his music. More often he uses the more disillusioned minor sixth-fifth – A flat-G in the key of C. *The Damnation of Faust* has not gone six bars in a serene D major before a wincing B flat has appeared to question Faust's contemplation of the beauties of nature and to insinuate the lurking *ennui* of self-consciousness. The flattened sixth in a major key is the worm at the heart of Berlioz's romanticism. It is the enemy within the gates, as in Cassandra's warning cry (a rasping G flat against the B flat major of the chorus) to the Trojans who wish to bring the horse inside Troy. It is the hyper-sensitivity which paralyses action – see the haunting conclusion of *At the Cemetery*, one of the *Nuits d'Été* songs, where the poet against his will feels himself held fast by the outstretched arms of the dead. In the Fantastic we hear the flattened sixth in the sixteenth bar of the theme (Ex. 110(*b*)), where it is as if the frustration of the initial melody at E-F-E had to express itself in the greater anguish of G-A flat-G), and at many other points in the harmony and melody: for instance, at the reappearance of the *idée fixe* in the slow movement:

111

But its importance in this work is expressive, not structural; it is subordinate to the more youthfully ardent and innocent fourth-third (F-E) motif – as if the artist still believed it possible to realize the dreams which agitate him. The frustrations are not yet paralysing. But they are keen. It is significant that at one of the few points in the work where the restless rhythms and nervous, unsatisfied accents are stilled and a spirit of acceptance is momentarily achieved – in the latter part of the slow movement – the *idée fixe* appears without its anxious fourth:

112

which broadens still further into

113

(But a moment later we have this:

114

and not long afterwards the extraordinary loneliness of the cor anglais calling across the void, without the distant answering oboe heard at the beginning of the movement.)

The *idée fixe* (Ex. 110) is a unifying force in two ways. First, and most obviously, after being treated at length in the opening movement it appears in different guises in all subsequent movements. Half way through the Waltz it glides into focus, with a sudden downward modulation, still irregular in its phrase-lengths but softened in character, and for a moment, like an intimate conversation, takes the attention from the semiquavers of the dance, which recede into the background. In the slow movement it erupts on the calm of the landscape, first the head of the theme in the bass (see Ex. 111), and then in full on the woodwind, very unevenly stressed; later when a desolate tranquillity is restored, the head of the theme, without its poignant fourth (Ex. 112) is combined with the main subject of the movement in a sublime passage which has more than a touch of Beethoven in its hint of thirds in contrary motion and in its cadences with their sense of calm and space and the holy simplicity of their common chords. At the end of the March the *idée fixe* appears again, a pathetic reminiscence on the clarinet, truncated by a guillotine stroke of the full orchestra. In the finale it is repeated in full, but, as in a nightmare, changed into a vulgarized, cackling tune on the E flat clarinet, lewdly accompanied by rising bassoon arpeggios – deliberately commonplace now that its rhythmic stresses (whose importance is thus retrospectively emphasised) are smoothed out.

But it is not simply a matter of the repetition or transformation of the *idée fixe* at certain key points. The *idée fixe* unifies the symphony in a profound way by influencing its general melodic character. It does so both by means of its F-E motiv and by its general form. The shape of the *idée fixe* (Ex. 110) – aspiring upwards by tones and semitones, then subsiding again – is reflected, in a calmer form, in the shape of the theme of the slow movement. This is another long melody – in its full statement, fifteen bars of *Adagio* – and, again, it is at first unaccompanied except by detached pairs of quavers.

115

The last section of the *idée fixe* (the last eight bars of Ex. 110) is paralleled by the last section of the Adagio theme (in its full version):

116

and, again, by a passage near the end of the movement.

117

The *Adagio* theme is an independent melody with a character of its own, but it is related – organically related – to the *idée fixe*.

The connexion between the second half of the *idée fixe* (Ex. 110(*b*) et seq.) and the opening phrase of the theme of the *Largo* introduction is commonly remarked on. They are in fact almost identical.

118

But one can go further and remark that the *Largo* phrase is also related directly to the *first* half of the *idée fixe*, as is made clear when the phrase reappears, at the same pitch, in the relative major, at the 28th bar of the *Largo*. It is simply Ex. 110(*a*), the F-E fixation and the descent from it (reduced in this example to C major).

119 

Throughout the work the F-E motif is continually active. The temptation to read in subtlety where none exists is always dangerous; but there are all sorts of places, too numerous to cite, where F-E, though not significant or even audible to the ear, is at work (including the *Dies Irae* in the finale and even the strange passage in ascending and descending chromatic scales in the first movement, which are linked to the *idée fixe* by the wailing semitone intervals, each a step higher, on woodwind and horns); and quite apart from F-E the symphony, as Schumann noted, teems with cross-references. Whether or not Berlioz was consciously aware of these things is not important to this discussion. The point is that they are there and that, partly because of them, the Fantastic is a work of marked thematic unity. None of this would make it a great work without the spirit which infuses the form. But spirit has never been seriously denied to the work; it is necessary to insist on the form.

Having established the thematic unity of Berlioz's first symphony we can proceed to examine his highly individual method of composing a symphonic movement, and also his application of the Beethovenian principle of drama embodied in musical terms in the 'Dramatic' symphony on *Romeo and Juliet*. Berlioz was perfectly capable of constructing a movement out of a short melodic figure – he did so, for example, in Herod's *scena* in *The Childhood of Christ* – but this was not the method he preferred. From the 'Fantastic' symphony to *The Trojans* he was primarily interested in extended melody. In this he was exceptional for his time and for some time afterwards. Wagnerian harmonic polyphony was to dominate music for many years to come. It is not easy even now to get the feel of Berlioz's melodies; their length, irregularity and harmonic independence often defies immediate acquaintance. Berlioz frequently starts with large full-blown melodic statements. The *Symphonie funèbre et triomphale* (1840)– a noble work which is only prevented from becoming popular by the very large wind and brass forces for which it was commissioned – begins at once with a wide-spanned melodic paragraph.

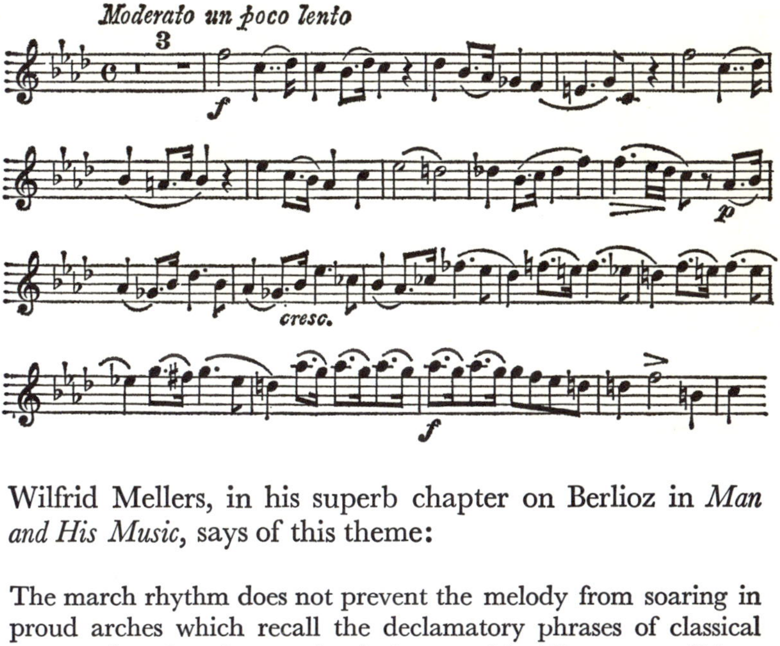

120

Wilfrid Mellers, in his superb chapter on Berlioz in *Man and His Music*, says of this theme:

> The march rhythm does not prevent the melody from soaring in proud arches which recall the declamatory phrases of classical opera rather than the symphonic themes of the Viennese tradition; and the asymmetry of the clauses is complemented by a tonal precariousness created by chromatic intrusions in the melody . . . Berlioz's chromaticism, unlike Wagner's, is almost always melodic, not harmonic; it extends, rather than disrupts, the span of the theme. Even the sequences which sometimes appear in his melodies are seldom strictly reflected in the harmony. Wagner's grandeur is the apotheosis of the personal. Berlioz thinks melodically in vast phrases that acquire a more than personal grandeur.

The first movement shows the maturer Berlioz now going more deeply into the harmonic implications of his themes.

The *Andante malincolico* of *Romeo and Juliet* (the introduction to the first movement proper) opens with a long violin *cantilena* (again largely unaccompanied) and then blossoms out into this controlled, impassioned melody:

121

Having proposed his melodic paragraphs Berlioz generally proceeds to construct from them not by fragmentation but by continual developing variation. The melody itself generates form. It is a method different from that of sonata form, but not necessarily less cohesive. The slow movement of the 'Fantastic' is a good example of this (a movement, incidentally, which took Berlioz much working and reworking over a period of two years before he was satisfied with it). The *Allegro* of the first movement is an even more striking illustration of Berlioz's method and its fundamental dissimilarity from sonata form. In a sense the 'development' has already been accomplished by the end of the forty bars of the *idée fixe*, in the generation of the complex sentences of the theme from the first phrase, with its two upward springs of a fourth and a sixth arrested by the F-E fixation. There is no question of contrasting the *idée fixe* with new material and achieving unity through tension of opposites. The subsidiary ideas are almost all offshoots and extensions of the theme, new light on the *idée fixe*, e.g.:

122

Only Ex. 123 is arguably new, and even it is related to the *idée fixe*, out of which it first emerges:

123

As for tonal contrasts, they do not have the same structural function that they have in sonata form. From the end of the exposition onwards the movement is a continually developing organism; there is no 'recapitulation' (even in the free sense in which Haydn understood it in his later symphonies) with the theme and its dependents processing through the tonic key; and there is no development in the Germanic sense.*

The whole course of the movement is concerned with the consummation of the *idée fixe* as a theme fully integrated with its harmony and accompaniment. At the outset, as we saw, the *idée fixe* is virtually unaccompanied, except for those isolated stabbing quaver figures which only emphasise its independence. (Berlioz's earlier idea – again discarded – was to have this first statement accompanied with block harmony on *tremolando* strings!) Half way through (Eulenberg miniature score, p. 28) the theme and its accompaniment have come closer together: the quavers are now continuous, and the violas add an extra part. (Here another confusing earlier idea, involving rising and falling scales on violas and 'cellos, was rejected.) Then a magnificent passage, in which

* Originally there was; the autograph score preserves (in what would be the development section of the present symphony if it had one) twenty-five cancelled bars, – the beginning of a passage of unknown length – consisting of this sort of thing, developed from Ex. 15 and rightly suppressed by Berlioz:

the head of the theme is treated in two-part imitation by violas and first 'cellos while clarinets and bassoons play syncopated chords and a soaring oboe line, derived from the theme, climbs higher and higher above them, leads to a *fortissimo* statement of the *idée fixe*, still with its stresses and syncopations, but now harmonized on full orchestra. Finally, the tempo slackens and the *idée fixe*, played quietly and with a less frenzied accent on the F, resolves itself in sustained chords – the theme has come to rest (temporarily) on the broad harmonies which it so proudly disdained at the outset of the movement. 'The consolations of religion' did well enough for a description, when Berlioz came to amend his programme to take in this revised (1832) ending of the first movement; but it is a programmatic description of an essentially musical process. 'Music is always in command.'

Berlioz's second symphony, *Harold in Italy* (1834), is also built round a single theme; but unlike the *idée fixe* in the Fantastic, this theme does not transform its character but remains always itself, the fixed point of reference for the changing scenes through which the Byronic hero passes. But the Harold theme has the same function as the *idée fixe* in generating thematic material for the work as a whole. One sometimes gets the impression from what has been written about *Harold* that Berlioz's method of unifying the work is simply to attach his motto theme more or less loosely at certain points to a number of otherwise unconnected movements. But it does not take a very close analysis of the work to make one realize that most of its tunes spring from one or more of the component phrases (*a*, *b*, *c*) of the melody which is foreshadowed (in the minor) in wailing woodwind octaves at the beginning of the work and then stated fully by the solo viola:

124

Even the music of the orgiastic finale, which brushes the doubting viola aside quite early on in the proceedings, is linked thematically and rhythmically with what has gone before. Harold, in fact, is not simply the observer of so many objective and picturesque scenes of Italian life; everything we hear is coloured by his mind, his poetic awareness, exuberance, introspection, anxiety. Even the objective fact of the distant bell in the Pilgrims' March monotonously tolling its C natural (the flattened sixth of E major, the movement's key) across the wide landscape of the Roman Campagna turns into an echo of Harold's private melancholy, just as the dwindling of the pilgrim procession and the gradual change from evening to night (marked in the texture and dynamics of the music) become symbols of the curve of feeling in the human observer of the scene, from contentment to *angst* and isolation.

In Berlioz's third symphony, *Romeo and Juliet* (1839), the drama has become more explicit and more openly reflected in the form, but the form remains symphonic, for all its bold extension of the genre. It can be argued that the recent failure of his opera *Benvenuto Cellini* (which ended the hopes of an *entrée* to the Paris Opera by which he had set so much store) forced him against his will to cast the next dramatic work in concert form, from which confusion a hybrid resulted, fascinating and beautiful in its parts, incoherent and unsatisfactory as a whole. This is a possible argument; but it is rather the argument of one who looks at the work from without, from a somewhat nice notion of symphonic proprieties and, seeing the unusual attempt to absorb techniques properly belonging to opera or oratorio into the symphony, expects it to fail. Berlioz did, much later, contemplate writing an opera – a totally new work – on the play, and it was age and ill health that stopped him, not the existence of a 'dramatic symphony' on the same subject. The line between his dramatic works for the stage and his dramatic works for the concert platform was a fine one, as Robert Collet has said, but he knew where to draw it. The *Romeo* we have is not a concert work with an opera inside trying to get out. This, at least, is to state one side of the argument. The other, with which many will agree, is summed up by Wilfrid Mellers when he claims that 'the structure of the work as a whole is a curious, not entirely convincing compromise between symphonic and operatic techniques'. But the work stands on its own, *sui generis*; but it does stand. The design of *Romeo and Juliet* is quite logical. The choral prologue states the argument; the choral finale, commenting on the tragic meaning of events and resolving the conflicts in a hymn of reconciliation, completes the scheme. The finale does not, of course, exactly balance the prologue, where the old tale is foretold in accents of timeless ritual; by contrast, it is caught up in the intervening tragedy. It serves the double function of an epilogue and of a concluding movement which, like the finale of the Ninth symphony, extends the work by bringing the instrumental drama out into the open. In the body of the work Berlioz's treatment of the drama is symphonic (though voices are used, just enough to keep them before our mind, so that the big choral conclusion shall not seem disproportionate) and his apologia is a symphonist's rationalization: 'The very sublimity of this passion is perilous for a composer who attempted actually to describe it . . . hence the recourse to the language of instruments, a language richer, more varied and highly developed, and by its very vagueness incomparably more effective in this instance'.

A gushing socialite once declared to Berlioz, after hearing the Feast at the Capulets', that she could 'absolutely see Romeo driving up in his gig'. But 'depiction' is the reverse of what he is aiming at in most of the work. The symphony, he told a friend, would 'make known to you the sum of passion that is in the play'. Berlioz, in Bernard van Dieren's words, sucks all the music from the verse. The Queen Mab scherzo, and the scherzetto which foreshadows it in the prologue, stand not only for Mercutio's speech but for the whole comic-fantastical element in the play, the fanciful nimble-witted spirit of mockery of Romeo and his coolly fiery companions. The Feast at the Capulets', with all its outward panache and brilliance, is a piece of psychological drama; the dance is not simply depicted but interpreted. It is the play's essence, not its detailed action, that is transmuted into symphony. During the Love Scene there is a passage where, in the quick palpitations of the woodwind and the cellos' pleading recitative, we seem to hear an echo of Juliet's agitation at discovering that Romeo has overheard her confession to the night, and the ardent voice of Romeo answering her from the shadows of the Capulets' garden. But we should not look for any closer parallel between the music and the balcony scene in Shakespeare. There is, rather, a general similarity of mood, and of shape: just as in the play the lovers cannot bear to say goodnight but linger talking until dawn, inventing fresh reasons for not parting, so Berlioz marvellously draws out the end of the movement, varying with richer and richer harmony and orchestration the love theme and its offshoots.

The 'Dramatic' symphony is laid out as follows:

PART I. A rapid fugal introduction – representing the fury of the ancient vendetta between the rival houses of Montagu and Capulet, which has erupted again in pitched battles in the streets of Verona – is followed by a solemn passage for trombones (the fugue subject in double augmentation) representing the intervention of the Prince; at its conclusion the fugal *allegro*, which has continued to growl in the background, returns, muted and fragmentary, and quickly withdraws to a distant mutter, making way for the Prologue. Solo contralto and semi-chorus now propose the drama: the upsurge of the feud, the Prince's interdict, the truce during which a brilliant ball is given by the head of the Capulets, the melancholy of Romeo (who is a member of the Montagu family), the ecstasy of the young lovers. Important themes developed later in the symphony are prefigured. Solo contralto, accompanied by harp and wind (and cello obligato in the second verse) sings of the passion of first love. Tenor solo and chorus mock Romeo's infatuation and declare that Mab, the spirit of dreams, has bewitched him. A brief concluding section describes how death holds sway and reconciles the chastened families.

PART 2. The first movement proper (Romeo alone – sadness – concert and ball – great festivities at the Capulets') is a symphonic *Andante-Allegro*. It opens with a long thematic passage played *pianissimo* by unaccompanied violins, suggestive of silence, suspense, breathless uncertainty, aspiration – the exalted feelings of Romeo alone in the Capulets' domain:

125

This passage – whose phrases alternately soar and pine, and whose melodic line walks on air above the common tonal earth – is the source of much of the subsequent material of the symphony. Its first act is to generate the full-blown theme already quoted earlier in this chapter, Ex. 121. A rhythmic figure, presage of the music of the Ball, interrupts the development of Ex. 121, and continues to punctuate the pure, soaring oboe tune (*larghetto*) which follows. This oboe tune is related, like Ex. 121, to Ex. 125.

126

Ex. 126 returns in the course of the succeeding *Allegro*, intoned by the woodwind and some of the brass against the main theme of the *Allegro* (Ex. 127) (itself a relative of Ex. 126) and the rhythmic figures that derive from it.

127

This *Allegro* is a continually developing movement; but it hardly ever leaves the tonic key of F major. Yet we are not aware of any limiting effect of this unvarying tonic, which deliberately creates a feeling of reckless gaiety, as the theme seems both to whirl forward and to remain in the same place – the heat and clamour of the ballroom expressed symphonically. Towards the end the pace (though not the tempo) grows increasingly hot and hectic with a series of string passages in triplets rising maniacally to brief explosions of brass and percussion. There is a moment just before the end when the scurry and tumult break off abruptly and the tapestry of sound parts as if to reveal the silent, watching Romeo, the oboe playing Ex. 126 above a descending bass (prominent earlier in the movement) while the drums continue to tap out the rhythm of the dance. Without doubt the piece is poetically suggestive in this sort of way; earlier, the stealthy *fugato* (a derivative of Ex. 127) first heard on the bassoons, may be seen as evocative of the gliding movement of masked dancers. But the poetic idea is thoroughly absorbed. 'Music is wholly in command.'

PART 3. Love Scene. The orchestra softly evokes the scene: the Capulets' garden on a hot summer night; low flutes and long string chords, with an occasional double bass note, *pizzicato*, vibrating into the stillness. A horn calls, a rocking monotonous call on two notes. Themes from the *Allegro* (Ex. 127 and its derivatives) are now heard, transformed, as distant snatches of song from an offstage semi-chorus of tenors and basses – the voices of the young Capulets returning from the ball:

128

Then silence falls, voices are banished, and the orchestra settles down to the long raptures of the Love Scene.* Out of

* As Jacques Barzun remarks in *Berlioz and the Romantic Century*, Shakespeare virtually tells Berlioz to compose his *Adagio* in pure sound:

How silver sweet sound lovers' tongues by night,
Like softest music to attending ears!

the rich pattern of phrases on the lower strings (divided into four parts), the wisps of melody on the violins and the sighs of cor anglais and clarinet, the great love theme which is the subject of the movement slowly flowers:

129 

This is the moment towards which everything has been growing, and in the radiant light of which the ensuing tragedy is seen. Perhaps of all Berlioz's music it is in the Love Scene that his obsession with the Romantic idea of a total, all-embracing love found fullest expression. The extraordinary purity of the music is not a paradox; it suggests not restraint but a passion so incandescent that it transfigures sensuality.

PART 4. For his scherzo Berlioz takes the Queen Mab speech and translates the flickering fantasy of Mercutio's images into a dreamlike yet glittering web of orchestral sound. The impression of darting movement is achieved by the harmony and whole style of the piece and by the frequent pauses – in which the music seems to hover, quivering intensely yet almost imperceptibly – as well as by the extremely rapid tempo and the delicate orchestration, with its air of moonlit transparency, luminous but unreal. The music has the character of some fantastic game, played in silent, gleeful concentration, between strings and woodwinds. The glassy, translucent music of the trio is based on the scherzo theme augmented. In the reprise of the scherzo, horn fanfares lead to a climax in which metrical irregularities and overlapping phrase-lengths intensify the atmosphere of eeriness and fantastical humour.

Originally Berlioz included at this point in the work a second prologue acquainting the listener with the background of events to the later scenes in the symphonic drama. Some time between the first performances and the publication of *Romeo* in 1847 he decided that it was an encumbrance and removed it.* In the movement which now follows, the Funeral Procession, the voices return and, while the orchestra plays a slow *fugato* the chorus chants a psalmody in octaves. The building of a movement round a single repeated note is a favourite device of Berlioz – see, for example, the Offertorium in the Requiem and the septet in the fourth act of *The Trojans;* (the use of the horns' reiterated C natural in the Pilgrims' March in *Harold* is a related procedure). Half way through this movement choir and orchestra exchange roles; the voices now dominate, in preparation for the choral finale. At the end the procession passes into the distance, leaving flutes and first violins sounding a bell-like dirge on repeated Es against alternating Es and Fs (and, after that, the successive notes of a descending modal scale) on the other strings. A second detached symphonic movement, Romeo at the Tomb of the Capulets, brings us nearer to pictorialism and prepares for the narrative finale. (It is based on Garrick's version of the play, in which Juliet wakes before the poison has killed Romeo.) The controlled frenzy of the opening suggests, perhaps, the slaying of Paris at the entrance of the tomb; it is followed by utter stillness – a series of string and wind chords, interspersed by long silences and seeming to listen for any sound. The cold, cavernous solemnity of the succeeding *Largo* (another magnificent piece of extended self-generating melody, scored for cor anglais, bassoons and horn, with displaced accents in the accompaniment, like a heavy heart beat) evokes the emotions of the dying Romeo alone in the tomb of Juliet and her ancestors. Juliet wakes to a pathetic rebirth of the love music – a phrase from the opening of the *Adagio* on the clarinet, struggling to form itself in the void. The lovers' joy is frenzied and brief. The musical material is

* A copyist's manuscript, containing a piano reduction of this second prologue and also of the original (longer) version of Friar Laurence's monologue has lately been discovered.

formed by the shattering and disintegration into fragments of themes from earlier movements. Ex. 127 becomes Ex. 130:

130

The hectic brightness of the orchestral texture gives an effect of light-headedness, nightmare and imminent collapse. (Berlioz uses similar means at the beginning of *The Trojans* to suggest the hysterical jubilation of the doomed inhabitants of Troy.) The sublime love theme (Ex. 129), torn apart by high *tremolando* violins, spins frantically round as if in the distorted memory of a dying brain, clutching at life. The movement ends in annihilation, with a violently discordant *fortissimo*, followed by a last echo of the love music and Romeo's reverie, and nothingness.

FINALE. The finale rounds off the scheme of the work. The Prologue's remote (and static) style is replaced by a deeply involved and continually developing piece of dramatic music which, while helping to frame the preceding mainly instrumental movements, is also the culmination of the tragic events distilled in them. After a stormy introduction, representing the arrival of the Montagus and Capulets at the graveyard, Friar Laurence steps forward. He explains (in symphonically developed recitative) the sequence of events which provoked the tragedy. The Montagus and Capulets start brawling over the bodies (to the music of the fugal Introduction), but Friar Laurence rebukes them and preaches forgiveness and peace, and foretells in a noble aria that Verona will one day take its fame from the young lovers who died victims of an insane vendetta. Slowly, to their astonishment, Montagus and Capulets find themselves forgetting their bitterness (the music gradually relaxing and expanding over a series of long pedal notes) until, led by Friar Laurence, they swear an oath of eternal reconciliation.

Whether *Romeo* 'works' or not cannot be proved; it is, perhaps, finally a question of personal temperament and mood, and also of individual performance. It is not a favourite work of lovers of formula and neat aesthetic categories. It is also a sensitive work; the atmosphere it enshrines is not crudely unlocked. One may well admire the movements separately but not *feel* that they make up a living entity in the concert hall. The dramatic point of some of the numbers can be obscured by dull and unstylish performance – for instance the concluding Sermon which easily assumes a jogging Meyerbeerian heartiness, but when taken at the very steady, dignified tempo indicated (♩. = 54) becomes a stirring expression of reconciliation and brotherhood. Given an answering sensitivity in the performers, it is possible to discover how carefully interdependent are the parts, how coherent and curiously convincing the whole. *Romeo and Juliet* is an extreme case; but its lesson is the same as that of the Fantastic. Far from being the works of a genius of uncertain aim and divided impulse, Berlioz's symphonies, when approached without preconception and studied with understanding, are revealed as the deeply thought utterances of a master of heterodox but precise form – their drama worked out in terms of controlled symphonic design, their most passionate feelings objectified into statements that transcend the ego which produced them. Taken for themselves, they are monuments to a poetic awareness and dramatic power which extended, without disrupting the flexible bonds of the symphony.

7
FELIX MENDELSSOHN
(1809 – 47)

Julius Harrison

*

As a composer of symphonies Mendelssohn is known to us chiefly through the 'Italian', and the 'Scotch'. The three others: 'Hymn of Praise' (*Lobgesang*); the early C minor composed in 1824, and the 'Reformation' have not found a comparable place in the repertory, though indeed the 'Hymn of Praise' – more a cantata than a symphony proper – has long since become a standard work for those smaller choral societies whose love of a good shout in rousing numbers has kept alive many a local body of singers unable to cope with the complexities of more modern music. As such, this 'Symphonie-Cantate', as Mendelssohn called it, has its rightful place, albeit rather outside the scope of the present article.

Before Mendelssohn was fifteen he had already composed twelve symphonies, eleven for strings and one for full orchestra. Most of them remain in manuscript. That he was a prodigy like the young Mozart is generally conceded. The list of his boyish compositions from 1820 to 1824 is remarkable. Sonatas, chamber-music, those twelve symphonies, many fugues, motets, two one-act comedy operas and much other music flowed from his pen in a display of precocity greater than Mozart's. In extemporization he had no equal. Goethe took the greatest interest in the boy from 1821 onwards, even venturing the opinion that whereas the very young Mozart had the 'prattle of a child' – Goethe had heard him in 1763 – Mendelssohn at twelve 'had ideas more grown-up and independent than were Mozart's at the same age'. Once, playing a Bach fugue, his memory failed him; yet he continued it with an extemporized ending in perfect accord with Bach's style. When the score of the *Midsummer*

Night's Dream overture was left in a cab by the English musician Thomas Attwood, and lost – it only turned up again in 1906 – Mendelssohn wrote out another which was found later to agree in every detail with the orchestral parts.

Apart from the C minor symphony – 'No. XIII' on the autograph score and known to us as No. I – the four that followed were either the result of commissions, or in the case of the 'Scotch' and 'Italian', the musical sequels to a Grand Tour of Europe planned by Mendelssohn's father and lasting no less than three years (1829-32). (Affluent indeed were the circumstances which enabled the Mendelssohn-Bartholdy family to indulge its passion for travel and sight-seeing when and where it would.) The *Lobgesang* (Symphony No. 2) was composed in 1840 as part of a Leipzig Festival commemorating the invention of printing by Gutenberg four centuries earlier; the 'Reformation' (No. 5) for the tercentenary of the Augsburg Conference of 1530 which defined once and for all the doctrinal tenets of the Lutheran Church as opposed to those of the Roman Catholics. Yet such was the commotion created by the Catholics in 1830 (that year of general turmoil in Europe) that Mendelssohn had perforce to abandon the project, finally contenting himself with a Berlin performance two years later. Under him the symphony was rehearsed in Paris that same year but never performed. The orchestra found it 'too learned', with 'too much fugato', and 'too little melody' for the taste of a city given over largely to *opéra bouffe*, revues and other entertainments exceeding in improprieties anything ever romanticized by Murger in his *Vie de Bohème*. After that experience Mendelssohn left Paris never to return.

But he loved England, and Scotland and its peoples. For over twenty years he was almost as much in these countries as out of them. Even London's smoke and fog had their place in his affections. He became *persona grata* with the queen and prince consort; the 'Scotch' symphony was dedicated to her. With that work and the *Hebrides* overture it can be said that he put Great Britain on the world's map of music, if in no more than a geographical sense. Both works are pictorial landmarks; both tell us that when he was stirred by nature's provenance or by some historical or local event, Mendelssohn's musical imagination rose to the occasion to produce a masterly composition reflecting the spirit of what he had heard and seen. Briefly then, he was a genius at orchestration, and nowhere more convincingly so than in the works just mentioned, in the *Midsummer Night's Dream* music and the 'Italian' symphony.

From his earliest years he had an instinctive feeling for what instruments could do effectively. His scoring of the C minor symphony at fifteen had already indicated how neat, even fastidious, and colourful each detail must be in its expression, to say nothing of the balance between the various instruments necessary to ensure a proper distribution of the sounds in the bigger passages. His great octet for strings, composed less than a year after this symphony, proved his mastery over complicated contrapuntal music. At seventeen he had composed the *Midsummer Night's Dream* overture with its gossamer-like passages for strings, its perfect woodwind writing, its humorous touches, its adroit inclusion of the now obsolete ophicleide* to illustrate Bottom's asinine 'translation', and his use of violins to imitate those raucous noises without which any donkey worth his carrot would be but poorly equipped. At a time when Berlioz was flaunting the trombones in many extraordinary passages as thrilling as they were often satanic in effect, Mendelssohn would argue that these instruments were 'too sacred to be over-used'. Throughout the four symphonies (the 'Hymn of Praise' excluded) they are only to be found in the 'Reformation'. Double woodwind, two each of horns and trumpets, timpani and strings suffice for the 'Italian' symphony; two extra horns colour more darkly

* Superseded by the bass tuba. 'In the forties,' wrote Adam Carse in his *Orchestra from Beethoven to Berlioz* (Heffer), 'all the large German orchestras had their tuba, and the serpents, bass-horns, and what ophicleides there were, retired into the lumber-rooms and museums.' Mendelssohn used a 'Contrafagotto e Serpente' in the finale of the 'Reformation' symphony.

much of the Scottish, the more sombre aspects of his subject seeming to require just that weightier differentiation in orchestral expression. Both these symphonies are in the key of A; movements in A major and minor are common to both, yet with what a difference! The A minor *Saltarello* which ends the Italian has the invigorating quality of a wild dance under a blazing Roman sun; in the 'Scotch' the A minor section of the finale is fiercely clan-warlike in mood – *Allegro guerriero* on the symphony's title-page aptly describes it. Mendelssohn's use of the timpani is lavish, but always to the point. His writing for the horns is often magical, as for instance in the third movement of the 'Italian' symphony. In sum, he regarded each instrument as a live personality; 'the pleasant intercourse with the old familiar oboes and violas and the rest, who live so much longer than we do, and are such faithful friends' is how he himself put it.

Of the C minor symphony (No. XIII/I) little can be mentioned here. All the same it is a remarkable work needing no apologia because it was the work of a boy of fifteen. Classical throughout in form, its first movement, *Allegro di molto*, and its finale, *Allegro con fuoco*, have a C minor vitality and restless urge foreshadowing (if not the equal of) the *Ruy Blas* overture of 1839. In the tender *Andante* movement there are beautiful *cantilena* passages for strings and woodwind, and one striking dramatic outburst to remind us that Mendelssohn was already at this early age a mature, self-reliant composer. The *Menuetto* movement, again *Allegro molto* and in 6/4 time, is truly Mozartian in its skilled counterpoints. Its trio, however, is all too static harmonically. Perhaps it was for this reason that when in 1829 Mendelssohn, seated at the piano, conducted the symphony for the London Philharmonic Society (to whom it was dedicated) he substituted for this movement an orchestrated, abbreviated version of the scherzo from his octet – for us today a precious addition to the repertory; music light as thistledown, in Tovey's words, 'quite as wonderful as the original'.

We next come, chronologically, to the 'Reformation' symphony, begun in 1828, and, as already stated, first performed four years later in Berlin. It seems strange on the face of it that Mendelssohn, born of Jewish stock, should have composed a symphony on such a subject avowedly Lutheran. But during the French occupation of Hamburg life had become intolerable for Jewish families. And so Abraham, Felix's father, took the step of having his children baptized into the Protestant Christian faith – a conversion that was to determine so much in Mendelssohn's adult life, both musically and in other ways.

The work (in D minor) consists of four movements. A religious *Andante* introduction in which is twice quoted the 'Dresden' Amen said to have been composed by Johann Naumann (1741-1801) and which Wagner was later to use so eloquently in *Parsifal*. An *Allegro con fuoco* movement in D minor follows without a break. Its first notes have the same rhythmical shape as those in Haydn's last 'London' symphony (No. 104) and Schumann's C major symphony.

131

The whole movement, strenuous and unyielding apart from one further reference to the 'Dresden' Amen, seems to illustrate the bitter spirit of religious warfare. It ends on a note of defiance with Ex. 131 having the last word. The two middle movements have next to nothing to do with the symphony's primary object. In the B flat *Allegro vivace* movement we can detect the buoyant spirit of a Haydn minuet. And then its G major trio section (oboes in thirds, flute trills, etc.) is a delicate parallel to the third movement of the 'Italian'. After the return of the B flat music, an exquisite coda is added that steals away on tiptoe. The G minor *Andante* movement might well be styled a 'Song without words'; it is little more than a brief interlude for flutes, bassoons and strings, to which in the closing bars (other instruments added) a sudden violent reference is made to the second main theme of the first movement. Strife is given its *coup de grâce* here.

132

Then a held G on 'cellos and basses links the movements to the finale. This opens with a woodwind and brass setting of Luther's chorale *Ein' feste Burg ist unser Gott.* Violas and 'cellos help to enrich the texture as the setting proceeds to an *Allegro vivace* in 6/8 time which deals with casual references to the first phrase of the chorale. The main *Allegro maestoso* section now follows, the D major key firmly re-established. In the *fugato* development we can trace the source of those vocal entries Mendelssohn was later to introduce (often perfunctorily) into his oratorio choruses. The middle and final phrases of the chorale reappear, mostly on woodwind. And soon the trombones get a step or two nearer to the climax when, supported by other instruments and against the *fugato* elements, they thunder out its two opening phrases. Finally, a *Più animato* coda is reached in which, in unadorned, bare semibreves, the same two phrases of *Ein' feste Burg* conclude a symphony that in its outer movements sounds so often like the genesis of an oratorio.

The 'Scotch' symphony (No. 3), completed in 1842, illustrates the near-scenic aspect of Mendelssohn's romantic art. But it would be wrong to call it programme music. Like the 'Italian' symphony and Beethoven's 'Pastoral', it is 'more an expression of feeling than of painting', of impressions gained through travel and experiences. The first sixteen bars were sketched in 1829 when Mendelssohn visited Scotland, and, *inter alia*, heard and set down in his *Hebrides* overture, sound rather more onomatopoeic than any ever imagined for his 'Scotch' symphony. In Rome, while engaged on the 'Italian' symphony, he still was thinking of the other. There is, indeed, a close kinship between the A minor passage near the end of the *Saltarello* and the melodic outlines and harmonies used in the A minor *Andante* sections in the first movement of the Scottish. He had sketched those sixteen bars in Edinburgh at Holyrood Palace, contemplating the chapel where Mary Queen of Scots was crowned; 'now open to the sky, surrounded with grass and ivy, and everything ruined and decayed' . . . 'and I think that I found there the beginning of my 'Scotch' symphony'.

That beginning ripened during the twelve intervening years before the symphony's completion into the impressive *Andante con moto* first section. Sixty-three bars long, it instantly commands the attention with the opening theme, gravely, even dourly scored for woodwind, horns and the lower strings.

133 

Then the music, as authentic in its symphonic development as anything similarly placed in a Beethoven symphony, rises to dramatic heights before subsiding into a part-repetition of the opening theme, coupled to some of the semiquaver passages heard in the development. This leads to the main *Allegro agitato* section, its hushed first theme scored for clarinet and strings alone.

134 

The inevitable, finely-wrought, climax now arrives. Out of it, in temporary respite, the second main theme emerges, shared imaginatively between strings and woodwind.

135 

The development section is Mendelssohn at his finest. From its mysterious start, remarkable in harmonic modulations, he proceeds through many varying moods: warlike, tenderly sorrowful, and always agitated. Later the return of these mysterious modulations (starting from another key) provokes a veritable storm. Was Mendelssohn intentionally programmatic here? Who can say? But the storm episode certainly results in a magnificent *fff* climax, lifting the whole movement sky-high before it reaches the return of the *Andante* theme. There its last eight bars round off a movement that grips the mind throughout by virtue of its imaginative writing and its superb workmanship.

Close on its heels comes the *Vivace non troppo* movement in F major. In effect it is an engaging *scherzo* bubbling over with semiquavers and founded on a clarinet theme with a touch of 'Charlie is my darling' about its dotted quavers – something Mendelssohn may have remembered and set down.

136

Given such an opportunity for a bustle of semiquavers and with so many dotted notes continually enlivening the rhythms, Mendelssohn does not disappoint. Here, in 2/4 time he writes as fancifully as he does in the 3/8 *Midsummer Night's Dream* scherzo, composed much about the same time. Inspired counterpoints, translucent orchestration, and a second subject which after its *pp* entry breaks out into an ecstatic climax, all proclaim the composer's mastery. Finally, like the *Dream* scherzo and that of the octet, this *Vivace* movement steals away as noiselessly as fairies' footsteps at first light.

Nine introductory bars, somewhat forceful in style, now take the symphony to the *Adagio* movement in A major with its sentimental first theme, its over-sweet major thirds (marked x) on the strong beats of the bar, and its feminine cadences:

137 

The movement is mainly a series of embellishments of Ex. 137, made a little less cloying because the orchestration is so varied and colourful. After this theme has run its smooth course of twenty-four bars – its codetta like a passage in Beethoven's 'Harp' quartet – Mendelssohn now strengthens the fabric of the music by introducing a semi-martial theme starting in A minor, the rhythmical shape of which had already been sounded by horns in the nine introductory bars. Interspersed three times between the embellishments of Ex. 137 and always in a different key, this theme seems to suggest the quasi-pictorial nature of the movement. Near the end, the broken phrases of Ex. 137 (clarinet), punctuated by the timpanist's *pp* references to the martial rhythms, sound rather like the warrior's sad farewell. As Hubert Foss has pointed out, Mendelssohn and Tennyson had much in common. Both had their place in a comfortable Victorian society.

After this surfeit of major thirds, armchair harmonies and feminine cadences, the *Allegro vivacissimo* finale brings us back to the stern mood of the key of A minor. Many pictorial interpretations have been foisted on it: a wild highland fling; a gathering of the chiefs of clans and their retainers; the savage play of claymore and dirk, and so on: for the enjoyment of this mostly fine movement, we cannot do better than listen to it as absolute music, disregarding analogies and remembering only the rousing *guerriero* spirit of it all. There are actually five easily recognizable themes, developed by Mendessohn with great contrapuntal skill, and of which only the opening one (Ex. 138) and the second most important one (Ex. 139) can be given here.

138

139

The first (Ex. 138), warlike though it is, is curiously restrained in its dynamics through an unexpected, timorous *piano* at the fifth note – a feature persisting on its every appearance, except near the end of the movement. The second (Ex. 139) is plaintive in mood, particularly when finally transferred from oboes to clarinets and bassoon in a beautiful coda which leads to the *Allegro maestoso assai* ending of the symphony. Here, in A major, Mendelssohn adopts what is so well described (once again in Hubert Foss's words) as 'that high moral tone which was the first lesson which parents gave their children'. This stirring finale asks for more than the conventional, puritanical homily Mendelssohn gives us; it needed a coda of the victorious kind such as crowns the end of Brahms's C minor symphony. Queen Victoria and her entourage would no doubt have approved Mendelssohn's moral uplift here. But for us of today does it not sound uncomfortably like the sanctimonious welcome home of the errant son, contrite, forgiven, and embraced?

The 'Italian' symphony, the most popular of the five, the most extrovert, full of sunshine, joyous rhythms and inspired orchestral touches, caused Mendelssohn, strange to relate, many a heartache. In the first flush of its composition in 1831 he could write to his sisters 'it will be the gayest thing I have yet done, especially the last movement'; a year later 'it was costing him the bitterest moments he ever endured'. This heartache made him withhold the publication of the score all his life, and all because he was dissatisfied with that last movement, the *Saltarello*! To this day no one has yet discovered where the flaws in this wild dance of the Romagna exist, so perfectly constructed, imagined and orchestrated is the music in every detail.

When in Rome in 1831 he was working on this symphony concurrently with the 'Scotch'. If he found the 'Italian' came more readily to his pen – the 'Scotch' remained unfinished till 1842 – it was because he was enjoying everything Rome and Naples had to offer. March 1833 saw its completion, the happy result of all those impressions gained by visits to famous places, museums, art galleries, St Peter's and the Vatican with their ceremonials and their music, and to that *al fresco* religious procession witnessed in Naples (or Rome) which inspired the second movement, the so-called Pilgrims' March.

The first movement overflows with happy ideas. Against rapidly repeated woodwind chords the violins, in octaves, set the thematic impetus going at a quaver-pace which rarely slackens throughout the entire movement.

140 

And if this theme does rely on major thirds on the strong beats of the bar, yet sentimentality is discounted by the rhythmic vigour of it all. Indeed, the whole symphony continually avoids those weaker elements in Mendelssohn's music – feminine cadences, etc. – which at a later date were to pervade and even disfigure so many of his compositions. But to give a detailed description of this movement with its brilliant and beautiful orchestral touches is beyond the limits of the present article. The second main theme is introduced on clarinets and bassoons against quaver movement on the strings:

141

The third theme occurs near the start of the development section. In unorthodox style, it is a novel and highly successful departure from classical precedent and plays a great part in the rest of the movement after its *fugato* introduction on the second violins.

142

Later it returns in A minor – the 'Scotch' symphony key which Mendelssohn seemed unable to expunge from his consciousness while engaged on both symphonies. But as we would expect, all ends exuberantly in the major.

In the second movement, with its religious procession picture as a background, the loud introductory phrase sounds a solemn note. Heard three times during the course of the music, it heralds a beautiful theme for oboe, bassoon and violas,

143

which, when taken up by violins, is decorated by flutes in one of Mendelssohn's loveliest inspirations. Two other themes follow; the first on violins derived from (*a*) in the introduction; the second, in A major, on clarinets. Eventually the whole movement fades away to nothing; the procession has passed out of sight and earshot.

Somewhat in the style of a minuet and trio, the *Con*

moto moderato movement has unpretentious charm and instrumental contrasts of colour which make it one of Mendelssohn's most ear-haunting contributions to the gentler side of symphonic art. The first section is given to the strings with occasional support from the horns and scattered phrases from the woodwind that towards the end play a more important role. Trumpets and timpani are silent here. The middle section with its hunting horn phrases (bassoons added) and its ascending scale passages for violins (then flute)

144

takes the imagination to some enchanted Athenian wood where Titania's fairies 'creep into acorn cups and hide them there'. In the second part of this trio-section, trumpets and timpani add to the magic of music no one can resist. After the customary *da capo* repeat, the magical elements – horns, trumpets, timpani, woodwind, etc. – return in hushed phrases bringing sleep to mortals and fairies alike in our imagined Athenian wood.

The *Saltarello* gets realistically down to Roman earth with three main themes. Two of them, the 'jumping' and 'hopping' themes, belong to the dance proper; the third is a *tarantella* which Mendelssohn develops with great contrapuntal skill as the music whirls along breathlessly. This final movement of the symphony is indeed remarkable for the fact that it is written in A *minor*. Many minor-key symphonies eventually reach their summation in the major – e.g. Beethoven's Fifth and Ninth symphonies. But here in this A major symphony Mendelssohn reverses the process – and very successfully. For he takes us enthusiastically to the pleasures of the common folk, whose wild capers turn his A minor key into something close to the spirit of the major.

After an emphatic full-orchestra opening, the first 'jumping' theme is given to flutes in thirds:

145

the 'hopping' second to violins.

146

The *tarantella*, flowing with continuous quavers, is the main part of Mendelssohn's inevitable *fugato* development, starting with an extended passage for strings alone. All the other instruments gradually join in to produce thrilling climaxes. Then, near the close, the dance loses impetus, exhausted maybe with its own exuberance in an A minor coda which, as stated elsewhere, recalls the mood of the 'Scotch' symphony. But in the final bars the music recovers its unquenchable spirit, loyal to the A minor key.

The 'Italian' symphony is an outstanding example of its composer's youthful genius. Happy in its outlook, pictorial in ideas and orchestration, never sickly in sentiment, it remains one of those symphonies to which we return again and again to find enjoyment in every note.

8

ROBERT SCHUMANN

(1810 – 56)

Julius Harrison

*

ALTHOUGH during the long years of their courtship the course of true love never did run smooth for Robert Schumann and Clara Wieck, yet when the bitter opposition of her father was finally overcome and they were able to marry in 1840, this happy sequel had an effect on Robert's music as remarkable as it was profound. Up to that year he had composed many major works for piano solo; after it, not one of like calibre. Now his mind travelled along new paths. With more love in his heart than could be adequately expressed through a medium deliberately set aside, he turned instead to song-writing – some 120, including the *Dichterliebe* and *Frauenlieben und Leben* cycles, were all composed in that year of their marriage – and then to chamber-music and symphonies from 1841 onwards. But already in 1832 he had tried his hand at a G minor symphony, the first movement of which was performed by the local orchestra at Zwickau in Saxony (his birthplace) and the whole work, with its poorly revised scoring, at Leipzig a year later. Then the symphony was finally rejected by him as worthless. At the age of twenty-two orchestration had proved an art that was to remain something of a bugbear to him all his life, however much he managed in time to improve it. 'I often put in yellow where it should be blue' he wrote to a friend in the year of the G minor symphony.

But eight years later came his happy marriage and soon with it the renewed desire to create a symphony. No; *two* symphonies! February, 1841, saw the birth of No. 1 in B flat, known as the 'Spring' symphony; and then in the following September came the D minor symphony as a birthday surprise for Clara. This was revised in 1851, none too happily Brahms thought, and then called No. 4. No. 2 in C major was composed in 1845–6; No. 3, the 'Rhenish', in October 1850. Another, in C minor, was sketched in November 1841 and then abandoned in favour of the *Phantasie* in A minor destined to become the first movement of the piano concerto Clara was to play for the first time at a Gewandhaus concert in Leipzig on New Year's Day, 1846.

Such in general is the story of Schumann's symphonies. Unequal in their instrumental presentation, and too often failing (in their outer movements) to drive home many finely conceived ideas to their logical conclusions, they remain the most debatable of all symphonies composed by the great masters of the nineteenth century. Even so – if we can today borrow a poet's much abused line – well may it be said of each one of them as Cowper could say of England: 'With all thy faults, I love thee still'. Each contains many ideas and individual touches of genius enshrined in beauty; many dramatic moments to stir our emotional reflexes, as well as continual melodic and harmonic surprises and subtle play of rhythms in syncopated passages of a kind only Schumann could create from out of that restless mind of his. Here was his mortal handicap, which in little more than a decade after his marriage was to prove his undoing, even if through this very tragic fact he was able to create a vast amount of fine music pouring from his overcharged mind like a mad torrent rushing hither and thither. Whenever we hear the joyous, exuberant opening of the 'Rhenish' symphony or the superb imagery in its later Cologne Cathedral movement; or, again, the magical, playful charm of the *Romanze*-inspired trio in the D minor symphony's scherzo, and the all-too-brief yet tense drama which makes the introduction to the finale of this same work so moving an experience, then we know we are in the audible presence of a master, however many times he may elsewhere disappoint or try us with his excessive rhythmical reiterations, his uncertain orchestration and his ready-to-hand choice of quaver and semiquaver figurations often more suited to the piano than to stringed instruments.

Unlike Brahms, who let no external circumstances hold his mind in thrall and who continually sought the mastery of Bach and Beethoven, Schumann throughout his all too short life was a prey to many strange obsessions influencing his mind, and through them his music. Stimuli of the most romantic kind were as the breath of life to him. In literature he – son of a cultured bookseller as he was – turned avidly to the novels of Jean Paul Richter and to the fantastic tales of E. T. A. Hoffmann, his music continually reflecting the spirit of what, in a fever of excitement, he had learnt from them and from other authors similarly endowed. In a letter to Clara written in 1838 he confesses how affected he is by everything going on in the world: literature, politics and people; he longs to express his feelings and 'transmute them into music'. But an overspill of intense subjectivity of this kind rarely tends to make a symphony one of commanding stature. Yet in his failure to achieve such an end, Schumann certainly succeeds in winning our admiration in other ways, for each of his four symphonies is a genuinely inspired extension of those lyrical elements we find everywhere in his other instrumental music and his songs. The poet speaks in all Schumann does; his symphonies are in a class by themselves: their ideas, their mode of expression, even their orchestral miscalculations belong to him alone. As E. T. A. Hoffman wrote of 'Capellmeister Johannes Kreisler' (alias Robert Schumann), 'His friends averred that nature, in organizing him, had tried a new recipe and come to grief in its experiment, having admixed too little phlegm to his excessively irritable temper and his destructively flaming imagination.'

Schumann's orchestration of his symphonies has provoked a good deal of sharp criticism, not altogether undeserved. His woodwind scoring, always planned for the same eight instruments (pairs of flutes, oboes, clarinets, bassoons), is too much on the lines of a united family whose individual members but rarely dare venture out on their own. When they do, as for instance the oboe in the *Romanze* movement of the D minor symphony, the effect is really beautiful because this so seldom happens. Again, Schumann found it a problem to manage four horns in close harmony without cluttering up the sounds and so disturbing the general balance of the instrumentation in the more agitated *Allegro* movements. Throughout the C major symphony only two horns are employed, much to the gain of the music's clear expression. Yet (also on the credit side), Schumann with his emotional gift for lyrical melodies and sensitive harmonies knew how to limit these instruments to a single pair when writing his slow movements and scherzos (that of the E flat symphony excepted). His writing for trombones, particularly in solemnly harmonized or contrapuntal passages is full of imaginative touches; only in the 2/4 movements of the B flat and D minor symphonies does he overdo the blasting effect of their rhythmical punctuations when they double so much of what the four horns have already overdone. His writing for strings, when not made technically awkward under the fingers because of rapid figurations whose real home is the piano, contains many enduring things. Attempts have been made by various composers and conductors – Mahler particularly – to revise Schumann's orchestration. But to little purpose; for the essence of Schumann's warmly vibrant music resides in its forthright romantic appeal with all those personal traits, lovable characteristics and faults which, taken all together, made up the 'three-natured' man of the *Davidsbündler* – the impulsive, passionate 'Florestan' fighting the Philistines, 'Eusebius' the thoughtful critic, and 'Meister Raro' the restraining influence.

The source of inspiration for the First (B flat) symphony was a springtime poem by Adolph Böttger. To each movement Schumann added fanciful titles 'Spring's Awakening', 'Evening', 'Jolly Playmates', and 'Spring's Farewell', all of which he discarded before the symphony was published. Mendelssohn conducted its first performance on 31 March

1841, the whole work, sketches and full score, having been completed during the two previous months. In this happy work, ‘the unbroken continuity with which one burst of emotion succeeds another’ (Grove) illuminates the music with a special light all its own. No dark, melancholy thoughts intrude. The ‘Evening’ *Larghetto* is a fireside movement, one more descriptive of domestic bliss than any other symphonic movement ever written. In it Robert declares his love for Clara. The other movements, buoyant, exuberant, quixotic in the most original coda to the scherzo, tell us the whole romantic story of Robert Schumann the tone-poet. Even the steel pen with which the symphony was written he found, so he averred, lying on Beethoven’s grave in the Währing Cemetery, Vienna. More, it will be seen, went into the composition of this symphony than the mere assemblage of notes and rhythms.

Trumpets and horns begin the long introduction of thirty-eight bars with this theme (*a*) which in the *Allegro* becomes more rhythmically urgent (*b*).

147

Originally, the notes of (*a*) were the same as at (*b*) except for the different time-signatures. But the sixth and seventh notes of (*a*), then G and A, were both in volume and tone-quality at variance with the other notes owing to the peculiarities of the unvalved brass instruments of Schumann’s time, which he did not fully understand. Mendelssohn suggested the alteration as at (*a*). Schumann gladly agreed, because, as he said afterwards, the two hand-stopped notes (x) sounded as if they had caught cold!

As it moves towards the *Allegro*, the long introduction develops dramatically in the Beethoven manner. And if in it Schumann did indeed hear springtime sounds, then flute and clarinet might well in their brief exchanges suggest a pair of songbirds. The *Allegro* is founded on short lively phrases derived from Ex. 147(*b*), and is attended by semiquaver figurations which have a great bearing on the whole movement. The second main theme – clarinets and bassoons against viola semiquavers – brings placid contrast in its beautiful phrases. But the overriding mood is one of continual movement through all its impulsive, short-limbed phrases, brought to a halt though they are in one place by the dramatic intervention of Ex. 147(*a*), now in 2/4 time. Near the end of this vigorous *Allegro* movement a new theme suddenly breaks out on strings; it is another of those 'bursts of emotion' so typical of Schumann's plunges into the unexpected, imparting a warm glow to this very personal music before the movement ends in brilliant style.

Little description of the *Larghetto* further to what has already been written is needed here. Its serene melody in E flat – one such as Beethoven could have imagined and Brahms could well have had in mind when planning his own slow movements – sings its way through the music, first on violins, later on 'cellos, and then, in octaves, on oboe and horn. The strings have many elaborate passages in support. Episodes are few and never once disturb the general calm. Trombones are silent till twelve bars from the end, when in hushed tones – Ex. 148(*a*) – they foretell the scherzo's main theme – Ex. 148(*b*). No break occurs between these movements.

148

The lively G minor scherzo has two trios; the first in D major, the second in B flat. Here Schumann, by making the second trio entirely different from the first, went further afield than did Beethoven, who in the two trios of his Fourth and Seventh symphonies repeated the same music. A curious feature of Schumann's scherzo is his use of the third (bass) trombone to punctuate the bass of the harmony in the louder passages, the other two instruments being silent throughout. As a coda to the whole movement Schumann turns to broken phrases of the first trio and in varying *tempi – più lento, ritard.*, pauses, *quasi presto*, silences and so on. Fanciful, waywardly eloquent touches here remind us of Aldous Huxley in his *Music by Night*: 'Silence is an integral part of all good music.'

The finale returns to the vivacious mood of the first movement, starting off with an ascending scale passage for full orchestra which is developed variously throughout the movement.

149

It becomes dramatic, especially when taken up imitatively by the trombones against a background of string tremolos. But the main, very Schumannesque, impelling force in this finale comes from the theme in quavers which follows immediately after Ex. 149:

150

This theme, together with the secondary episodes growing out of it and Ex. 149, carries the main burden of the movement. Before its recapitulation on flute and bassoon there is a sudden lull: a *Poco adagio* cadenza for hunting horns and flute as unexpected as it is effective. The final *accelerando* coda to the symphony is both dramatic, classical in style, and immensely satisfying. Three timpani are used in this symphony. The triangle in the first movement is Schumann's only symphonic use of percussion.

In its general structure the C major symphony (No. 2) of 1845–6 is the most formal of the four. While, like the others, it still has its digressive moments, these owe their presence here more to Schumann's failure to live up to the full expression of his poetical nature than to his gift for inspired rhapsody. When the first movements were being sketched in December, 1845, he was suffering from one of those nervous breakdowns which were to result in his death ten years later. 'Only in the last movement did I begin to be myself again, and it was not till the end that I became completely well was how he described his condition at the time. Most neglected of the four, this symphony is actually Schumann's nearest approach to strict classical form and orchestral expression. Horns are reduced to one pair. Together with the trumpets they consistently avoid those chromatic harmonies without which Schumann must have felt like a lost soul, so great a part did they play in his music. Trombones are restricted to the two outer movements. And apart from the orchestration itself, Schumann was obviously at pains, in that spirit of 'obstinate conflict and determination' (Grove), to make his symphony conform largely to the diatonic harmonies of Beethoven's C major and minor movements. All the movements are pitched in C. The third, in C minor but ending in the major, recalls the *Marcia funebre* in Beethoven's *Eroica*; the fourth contains not only a facsimile of a phrase heard in Beethoven's *An die ferne Geliebte* and Mendelssohn's *Hymn of Praise*,

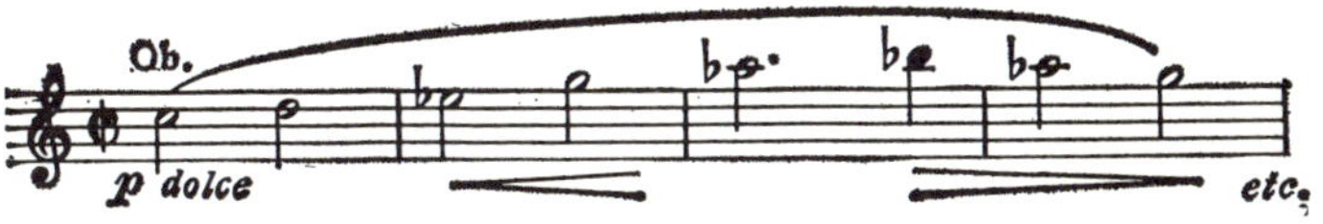

151

but also C major tonic-dominant figurations identical with those Schubert used in the first movement of his great C major symphony – a work Schumann discovered in Vienna after Schubert's death and heard for the first time under Mendelssohn in 1840.

Yet in spite of these unconscious plagiarisms, and the handicaps endured through ill-health, Schumann gave the world a symphony of some consequence. The *leitmotif* on trumpets, horn and trombone with which the symphony opens,

152

becomes a clarion call in the 3/4 section of this movement; is sounded again near the end of the scherzo, and then returns in the finale, hushed at first and then triumphant.*

In the *Sostenuto assai* part of the first movement, Schumann, with his characteristic sense of high drama, gives out many hints of themes and phrases to follow in the 3/4 section. There everything is consolidated and developed symphonically to complete a movement in which some rhythmical monotony 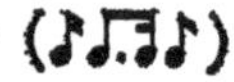 is countered by the general vitality of the whole.

There is much semiquaver bustle and feverish activity in the scherzo, which, like that of the B flat symphony, has two independent trios. The first trio is a delightful exchange of thematic phrases between woodwind and strings; the second a more tranquil movement starting on strings alone, from which (as in Trio I) all semiquavers have been banished except for the few which anticipate the return of the scherzo music. Schumann – in the guise of Florestan? – here scatters his ideas over a wide range of keys, using the diminished seventh chord as a pivot wherefrom to turn in any direction he pleases.†

The C minor *Adagio* movement opens with a beautiful and plaintive theme played by all the violins, and then, after

* See Ex. 131, p. 236.

† 'Clapham Junction is like the diminished seventh–susceptible of such enharmonic change, that you can resolve it into all the possible termini of music.' (*The Way of All Flesh* by Samuel Butler)

transference to the oboe in altered form (E flat major), is developed further in many moving passages.

153

The first four notes, both in this form and inverted, together with the poignant *fp* drop of the seventh in the fifth bar, return in the finale. There they are blended – a little haphazardly it must be said – with Ex. 151 and 652; with a passing reference to an important theme from the first movement:

154

and with many ascending scale passages founded on the finale's first bar.

In the finale the dark shadows of the *Adagio* vanish before Schumann's valiant determination to master every despairing mood. The Beethovenish *An die ferne Geliebte* theme (Ex. 151) which plays such a large part in this movement's progress may perhaps be connected with Clara.

In 1850 Schumann had become Director of Music at Düsseldorf and saw much of the Rhineland. The E flat symphony of that year – No. 3 but the last to be written – owed its inception, he declared, to the sight of Cologne Cathedral glistening in the sunlight; to the deep impression made on him by the grand ceremonial attending the installation of Archbishop von Geissel as Cardinal; and, in more popular style (fifth movement), to the outdoor merry-making of the Rhinelanders. It is therefore Schumann's nearest approach to pictorial symphonic music, though what he set down would seem to illustrate his pent-up emotional reactions to things seen and heard, more than anything of a definitely programmatic nature.

It we can judge by the invigorating character of the first movement, Schumann certainly took on a new lease of life when he wrote it. Its form is classical, and the music itself developed with inspiration and great skill. Trombones are withheld until the fourth and fifth movements. And if the four horns Schumann used tend to make some *tutti* passages too brassy, perhaps there is good reason for this, for the music is exuberantly happy and upsoaring, as if it were continually trying to reach the very spires of the cathedral itself. It opens with this striking and vivacious theme, that is to dominate most of the movement.

155 

In its spirit this movement recalls Beethoven's 'Eroica' once again; and could indeed have influenced Elgar when he composed the equally upsoaring opening to his E flat symphony. Other themes occur and are repeated in the usual classical manner. Yet all of them are subsidiary to the dominating Ex. 155. Schumann's build-up of the development section towards the *fff* recapitulation is one of his finest achievements. Had he lived ten years longer what might he not have done in the world of the symphony?

The scherzo is best described in Tovey's words: 'a slow *Ländler* with a comfortable Rhenish rusticity in its lilt'. Schumann discarded his original title for it, 'Morning on the Rhine'. It is a highly-organized movement in C major, having as its main theme one that is ingeniously varied with semiquavers – as in Ex. 156(*b*).

156 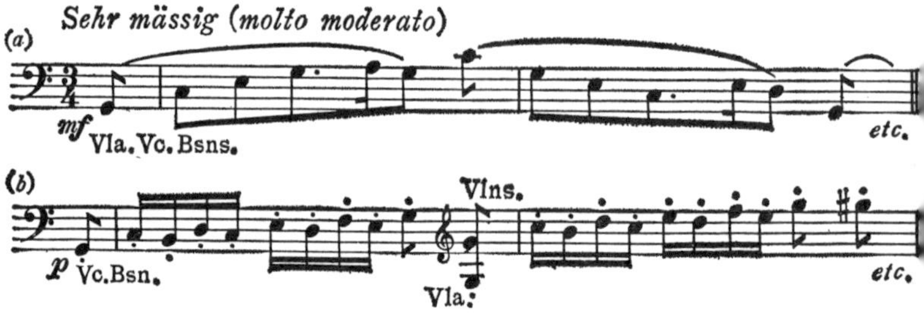

The original theme (*a*) is subject to richly-scored developments, marred at times by the excessive use of the four horns playing at a high pitch. The quiet coda, in which the 'cellos have their leisurely, gentle footsteps dogged by a cumbrous bassoon, echoes the movement's opening phrases. And if in this music Schumann did envisage a morning picture of the Rhine, then the *Nicht schnell* movement in A flat that follows certainly suggests a mid-day siesta, so unruffled is its calm. Trumpets and timpani are omitted; horns are reduced to two; and only five *fp* accents interrupt the pervasive *p*, *pp* and even *ppp* expression marks.

In the remarkable *Feierlich* fourth movement (E flat minor) the trombones enter for the first time in polyphonic music of extraordinary beauty and majesty. So inspired are the contrapuntal textures, gradually expanding from common time to 3/2 and then to 4/2, that it is no exaggeration to say that Schumann here exalted his genius to Bach-like heights of ecclesiastical expression and splendour. On the manuscript score he wrote *Im Character der Begleitung einer feierlichen Ceremonie* – of a character to accompany a solemn ceremonial. From the *pp* opening (after a *sforzando* first chord) an alto trombone, joined by the first horn, mounts to a high E flat in music rich in chromatic harmonies.

157

It sounds like a veritable *sursum corda* as the movement gradually unfolds dramatically through the 3/2 and 4/2 sections, aided as it is by much quaver movement on strings and woodwind derived from the first five notes of Ex. 157. Towards the close, two fanfare-like passages for wind and brass suggest the crowning point of all the solemn ritual. Then the music, still true to the E flat minor key, gradually subsides. Here in our mind's eye we can picture the congregation pouring out of the cathedral into the bustle of Rhineland holiday life.

The *Lebhaft* finale returns to the vivacious style of the first movement with a sprightly theme which has its offshoots in others of a like character.

158

Soon there are strong hints from four horns in unison that the polyphony of the cathedral movement is not to be forgotten. Here one can imagine that the holiday-making Rhinelanders found the grand ceremonial to be an exciting topic for conversation over many a *Stein* of their favourite brew. Then the quavers of the fourth movement also reappear to add further vivacity to the whole. Finally, a *fugato* version of Ex. 157 and oblique references to the first movement's opening theme (Ex. 155) help to build up an imposing ending full of *joie-de-vivre.*

The differences between the 1841 and 1851 versions of the D minor symphony (known as No. 4) are so many in the actual notes, notation and scoring that it is quite impossible to embark on any description of them here without exceeding the scope of the present article. For a detailed account the reader is referred to Tovey's article in his *Essays in Musical Analysis.**

There is in this symphony so much originality both in form and material, and so many transformations of the themes, that it can be regarded as a highly interesting example of free-and-easy cyclic form, so often does Schumann return throughout the entire work to the music of the first movement. Thus from the gravely beautiful theme of the slow introduction, Ex. 159(*a*), he evolves some part of the main subject of the *Lebhaft* section, Ex. 159(*b*), and in such a way that the five notes concerned (shown bracketed) are barely recognizable as belonging to the introduction.

* Oxford University Press.

159

From this point the music proceeds in unceasingly restless style: so restless indeed that the whole structure of the movement is heterodox and even vague. Yet at the same time the music is so true to Schumann's romantic, wayward temperament that we do not even notice how at a later stage in the movement he has contrived to avoid everything in the way of formal repetition, so skilfully does he build up the excitement through many dramatic pauses, mysterious string tremolos, and a whole series of chromatic sequences carrying the music forward to its ecstatic D major ending. One other phrase much used in the movement needs illustration, for with it Schumann starts the D major *Lebhaft* section of the finale.

160

With both these versions variants of Ex. 159(*b*) are combined in that spirit of close thematic integration which makes this symphony as interesting to study as it is enjoyable to hear.

Although the *Romanze* has for its main theme a plaintive A minor melody shared in octaves by oboe and 'cellos,

161

yet, paradoxically enough, Schumann devotes the greater part of the movement to variants of Ex. 159(*a*). These variants are dovetailed so naturally into the general flow of the music that he was obviously well content to let this oboe-'cello melody remain simple and undeveloped so as not to delay expressing himself still more deeply. In every way the *Romanze* is yet another example of Schumann's music of the homely, fireside kind. The variants of Ex. 159(*a*) are many: in A minor, and then, still more elaborately, in the longer D major middle section, where our now familiar phrase is charmingly decorated by a solo violinist's arabesques destined to play an even more entrancing part in the trio of the scherzo.

162

The oboe-'cello melody closes the *Romanze* movement. Its *pizzicato* accompaniment – here and at the beginning – Schumann originally intended should be played on a guitar – perhaps in the spirit of a birthday serenade for Clara, for whom the symphony was written.

The D minor scherzo makes a lively contrast to the *Romanze*, its main theme being given free canonic treatment – a feature persisting throughout much of the music here. In the rhythmical woodwind chords (flutes and clarinets) which act as part-accompaniment Tovey sees some resemblance to the shape of Ex. 160 in the first movement, though indeed the actual sounds heard would surely elude most listeners in this respect. The beautiful, sunny B flat major trio – see Ex. 162(*b*) – in which Schumann's imaginative powers are sustained at a very high level of inspiration, occurs twice. The second time it is subtly altered in playful, nostalgic retrospect as the music gradually dissolves into broken phrases. In a now overcast sky, clarinets, bassoons, violas and 'cellos sound like a portent, their reiterated harmonies leading without break into the *Langsam* introduction to the finale. There, for the first time in the symphony, the music is set in broad common time. In it Schumann creates a mood of strange foreboding, founded on phrases from the first movement – see Ex. 159(*b*) – against dramatic string tremolos and arresting passages for the brass. In his youth Brahms must have been deeply affected by these sounds, for he did much the same thing and even more dramatically so when in 1876 he composed the *Adagio* introduction to the finale of his C minor symphony. But Schumann's *Langsam* introduction is less extended. The clouds soon lift and the music of the *Lebhaft* section bursts into D major with the same degree of ecstatic emotion that marked the closing section of the first movement. From this point – see Ex. 160(*b*) – Schumann pursues his rapturous mood with various fresh themes, with *fugato* passages, dramatic silences and rhetorical passages for the brass. Finally, he reaches a *Presto* coda which, if all too brief considering the symphony's magnitude, and being, moreover, founded on a purely secondary passage from the previous *Schneller* (₵) section, brings this very personal symphony to its brilliant close.

Parry once stated: 'People who only write blameless symphonies and sonatas after the accepted models with nothing of their own to mark them by, might just as well let composing alone.' Schumann's are certainly not blameless; but they all contain those intensely human qualities which give them their rightful place in our affections and in the world's repertory of accepted masterpieces.

9
FRANZ LISZT
(1811–86)

Humphrey Searle

*

IT is difficult to write any general introduction to the 'Faust' and 'Dante' symphonies of Liszt. The two works are very different, and Liszt was not a symphonist in the ordinary sense – his achievements lay more in the direction of the symphonic poem. His only major 'abstract' work is the piano sonata in B minor which is in any case a succession of mood pictures rather than a work constructed according to strict classical principles. Apart from the early 'Revolutionary' symphony, which he later partly recast as the symphonic poem *Héroïde Funèbre*, Liszt wrote only two symphonies, both inspired by literary models. But these are not simply musical portrayals of the books that inspired them, with pictorial effects and all, like Strauss's *Don Quixote*, for instance; they are more in the nature of a musical commentary on a literary work; they express, in fact, Liszt's reactions to Goethe and Dante.

Formally they are far removed from the symphonies of Beethoven or even Berlioz. Liszt applied to them the principles he developed in his symphonic poems; the form depending on the subject matter to be expressed and with much use of 'transformation of themes'. Liszt's themes are short and plastic, and vary their character according to the mood of the moment; and vast edifices are built up from simple phrases. Liszt's use of the orchestral palette is also original and daring, and is generally calculated to make the utmost dramatic effect; and those who appreciate drama in music cannot really complain that the drama is overdone. Liszt showed great sensitivity in this direction, and his two works remain a remarkable, if unusual, contribution to the symphonic repertoire.

On 4 December 1830, the day before the first performance of the 'Fantastic' symphony, Berlioz relates in his *Memoirs*: 'I received a visit from Liszt, whom I had never yet seen. I spoke to him of Goethe's *Faust*, which he was obliged to confess he had not read, but about which he soon became as enthusiastic as myself.' Gèrard de Nerval's translation of *Faust* had appeared in 1827, and Berlioz had written his 'Eight Scenes from *Faust*' in the following year. In 1846 he produced his *Damnation of Faust*, which certainly influenced Liszt to write his own work; the 'Faust' symphony is in fact dedicated to Berlioz.

Liszt had some doubts about his task; shortly before embarking on the 'Faust' symphony he wrote to Princess Sayn-Wittgenstein: 'Anything to do with Goethe is dangerous for me to handle,' and fifteen years later he wrote to a friend: 'In my youth Faust seemed to me a decidedly bourgeois character. For that reason he becomes more varied, more complete, richer, more evocative (than Manfred) ... Faust's personality scatters and dissipates itself, he takes no action, lets himself be driven, hesitates, experiments, loses his way, considers, bargains, and is only interested in his own little happiness.' But though Liszt's attitude to Goethe was certainly not one of unbounded admiration, the Faust legend did in fact inspire him to produce two of his finest works, the 'Faust' symphony and the *Two Episodes from Lenau's Faust* (of which the second is the well-known first *Mephisto Waltz*), and there is very little doubt that the first movement of the 'Faust' symphony, with its constant and remarkable changes of mood, is a portrait not so much of Faust as of the composer himself.

Liszt called his work *A Faust Symphony in three character studies (after Goethe)*: 1. *Faust*; 2. *Gretchen*; 3. *Mephistopheles*. It occupies over 300 pages of full score, lasts more than an hour in performance, and was composed in the remarkably short space of two months (August to October 1854). He had, however, sketched out some of the themes in the 1840s, during his period as a travelling virtuoso, and the 1854 version differs considerably from the work as we know it today. It was a purely orchestral symphony without the choral ending, which was added three years later, and it was scored for strings, woodwind and horns only – there were no trumpets, trombones, harp or percussion. And it contained some experimental time-signatures; one of the themes from the first movement (Ex. 165 below) originally appeared in bars of 7/4 and even 7/8 at times.

Liszt tried the symphony out in private rehearsals at Weimar which were attended by many of his friends and colleagues, including Berlioz and Wagner, and he made numerous alterations after these. Up to that period, as a result of having to earn his living as a professional pianist, his study of orchestration had been somewhat sketchy, and a good deal of the scoring of the earlier symphonic poems had been carried out by Raff and Conradi. The 'Faust' symphony is one of the first works which Liszt completely scored himself, and no one can deny that the orchestration is, to say the least, extremely effective and striking.

The symphony was first performed on 5 December 1857, at Weimar, with the composer as conductor, in a concert in honour of the foundation of a memorial to Grand Duke Karl August (the patron of Goethe and Schiller) and also of the unveiling of monuments to Goethe, Schiller and Wieland.

However, Liszt still was not satisfied with it, and he again revised it extensively; the first performance of the new version was given at Weimar in 1861, and the score was published in the same year. Even after this he made a few small alterations, adding ten bars to the second movement as late as 1880.

The 'Faust' symphony is scored for normal symphony orchestra, with the addition of piccolo, third trumpet, harp, and (in the last movement) tenor solo, male chorus and organ. Apart from some passages in the second movement, Liszt does not attempt to portray Goethe's story in detail, and the first movement is a character study of Faust seen from many angles. Its form is difficult to analyse; it may be roughly described as consisting of a very long exposition followed by a short development section and a condensed recapitulation. It begins with a slow introduction which states two of the principal themes. The first, Ex. 163, with its whole-tone flavour, might be said to represent the mystical and magical element in Faust's nature. (Incidentally Wagner later made a considerable use of this theme in *Die Walküre.*)

163

It is immediately followed by the second, Ex. 164, a theme which assumes many aspects during the work, but generally represents Faust's emotional character, whether passionate, amorous, or melancholy.

164

The *Allegro impetuoso* which follows has a feeling of passionate striving, and leads to a *fortissimo* statement of Ex. 163 on the trumpets and trombones, ending with sharp dramatic chords. The second theme, Ex. 164, on solo bassoon introduces a new theme, Ex. 165, of a stormy character. Beginning on the strings alone, it rises to a passionate climax, and is restated by the full orchestra.

165

It leads straight into another emotional theme, Ex. 166, which plays an important part in the second movement.

166

The excitement then subsides, and a passage marked *misterioso e molto tranquillo* introduces Ex. 163 on clarinets and *pizzicato* strings under an undulating violin figure. This leads through a short *plintivo* section based on Ex. 164(*a*) to an expressive statement of Ex. 164 on woodwind and horns, answered by solo viola phrases, which plainly portrays Faust's amorous nature. As the tension increases, the tempo quickens, and horn and trumpet calls herald the last of the Faust themes, Ex. 167, representing his heroic aspirations.

167

Ex. 164 soon joins it, and the exposition ends with a *fff* restatement of Ex. 167 on the brass, accompanied by rushing string figures.

There is no break in the music, however, which dashes on into a restatement of Ex. 163 in canon between trombones and trumpets, followed by Ex. 165, which again rises to a powerful climax. Now at last the excitement dies down, and the whole of the slow introduction returns, followed by an *Andante mesto* section, in which Ex. 164 on clarinet and bassoon is combined with Ex. 163 on the lower strings, and Ex. 166 also makes a short reappearance. Now follows one of the most remarkable passages in the work; Ex. 163 is heard on clarinet and *pizzicato* violas under whole-tone chords for flutes and violins tremolando; all key-feeling disappears, and the effect is magical in the true sense of the word. Then a gradual *crescendo*, still based on Ex. 163, leads back to a violent restatement of Ex. 165, with which the reprise begins.

This section is much condensed; after Ex. 165 has reached its climax, Ex. 164 is heard again in its amorous form, and leads back to Ex. 167. This first appears quietly on solo trumpet, horns and bassoon, but a *crescendo* passage based on Ex. 164 soon leads to a *fortissimo* restatement, followed, as before, by Ex. 163 in canon on the brass. Finally Ex. 167 rings out again on the trombones, and leads to a short coda, mainly based on Ex. 163. At the end Ex. 164 breaks in abruptly; the music seems to collapse into its main tonality of C, and it is Ex. 164 in its original form on the lower strings which has the last word.

After the storm and stress of the Faust movement, which seems to have been written at white heat throughout, the atmosphere of the second movement, Gretchen, is of a magical delicacy. A short introduction on flutes and clarinets leads to the main Gretchen theme, Ex. 168, on the oboe accompanied by solo viola – a characteristically original piece of scoring, and exactly right as a musical portrait of fresh and innocent youth.

168

Soon a reference to Ex. 166 shows that Faust is not far from Gretchen's thoughts; it is followed by a short dialogue between clarinet and violins, depicting the scene where Gretchen plucks off the petals of a flower, murmuring to herself: 'He loves me – he loves me not – he loves me!' Then Ex. 168 returns with fuller scoring, and leads to the second main theme, Ex. 169, marked *dolce amoroso*.

169

As it dies away, Ex. 164 on the horns (marked *patetico*) indicates the entrance of Faust on the scene, and the mood becomes more agitated. It is followed by Ex. 166 on the 'cellos, *espressivo con intimo sentimento*, and finally the full orchestra enters with a quiet statement of Ex. 164 in its amorous form. The lovers have come together, and even the stormy theme, Ex. 165, can now be heard *soave con amore*. This leads to the reprise, which follows the exposition more or less exactly, except for the omission of the flower episode, and the insertion of Ex. 164 before the reappearance of Ex. 169. As the movement dies away, a quiet reference to Ex. 167 seems to show that Faust's happiness is now complete.

Mephistopheles is the spirit of negation; and to represent him Liszt hit on the ingenious idea of using only parodies of the Faust themes, with (as we shall see later) one exception. After a short introduction, marked *ironico*, which sets the malevolent atmosphere, Ex. 164 appears in the abrupt form heard at the end of the first movement, and soon there is a hint at a new figure, Ex. 170, which Liszt borrowed from an earlier work, the so-called *Malediction* concerto for piano and strings, where it is marked *orgueil* (pride) in the original score.

170

Next comes Ex. 163 on the clarinet accompanied by chromatic scales on the violas, followed by a rhythmically altered version of Ex. 165. It runs its course much as in the first movement, and works up to a powerful climax. Exs. 163 and 164 then reappear in their parodied forms, and a trill passage in the strings and woodwind leads to the appearance of Ex. 170 in the form quoted here.

After a false start Ex. 164 now starts up as a fugue on the strings, and afterwards its 'abrupt' form leads to a *fortissimo* statement of Ex. 167, which is parodied chiefly by the addition of trills. It is interrupted by Ex. 164 (followed by Ex. 163 *pizzicato* under stopped horn chords), but Ex. 170 again builds up a climax, at the crest of which Ex. 167 returns. Ex. 164 continues the savage and violent mood, till a sudden *pianissimo* heralds the Gretchen theme, Ex. 168 – the only one which is proof against parody.

But the respite is only momentary, and the diabolical dance begins again with Ex. 163, followed by a full statement of Ex. 165. Ex. 167 reappears, and the tempo increases; finally Ex. 164 carries the music on to the ultimate climax, crowned by Ex. 170. Twice Ex. 165 interrupts, followed by Ex. 163 and the stopped horn chords; but the onward surge continues until the crest is reached, and the music sinks down in exhaustion. Then the entry of the Gretchen theme, Ex. 168, on horn and solo 'cello leads to the final section of the movement.

This point was the original ending of the work, and it is still played in this form when no chorus is available, but there is no doubt that the final choral section sums up the whole work and makes it complete. First the trombones hint at the choral theme; then after a pause the male chorus enters, singing the 'Chorus Mysticus' which ends the second part of Goethe's *Faust*: Louis MacNeice made a brilliant translation of this almost untranslatable poem.

Alles Vergängliche	All that is past of us
Ist nur ein Gleichnis;	Was but reflected;
Das Unzulängliche	All that was lost in us
Hier wird's Ereignis;	Here is corrected;
Das Unbeschreibliche	All indescribables
Hier ist es getan;	Here we descry;
Das EwigWeibliche	Woman's divinity
Zieht uns hinan.	Leads us on high.

At the words 'Das EwigWeibliche' the tenor soloist enters with the Gretchen theme, Ex. 168. Higher and higher he soars, till eventually the whole passage returns *fortissimo* with solemn fanfares. Again the tenor takes the Gretchen theme up to the heights; and finally Ex. 167 in the bass brings the whole work to a triumphant conclusion. With all its faults and longueurs, there is no doubt that in it Liszt wrote his masterpiece.

Dante was one of Liszt's earliest loves; he had read him frequently in the 1830s with Marie d'Agoult, and in 1837 had written the first version of his *Après une Lecture du Dante* for piano (usually known as the 'Dante' sonata). By 1847 he had already sketched out the principal themes of the symphony, and intended at that time to get the painter Buonaventura Genelli to design lantern slides to be shown during the performance of the music. But he did not get to work seriously on it till the summer of 1855, and completed it in July of the following year. His original idea was to write three movements, Inferno, Purgatorio, Paradiso, corresponding to the three sections of Dante's *Divina Commedia* but Wagner persuaded him that no human being could express in music the joys of Paradise, and the work now ends with a Magnificat for women's voices in place of a third movement. This is a pity, for the balance of the symphony is thereby destroyed; and though Liszt was certainly more at home in the infernal than the celestial regions, the task should not have been beyond his powers.

The work is dedicated to Wagner, and was first performed under the composer's direction at Dresden in November 1857. It was a failure, owing to lack of rehearsal, but on its repetition in Prague the following year it scored a great success. An introduction to the work by Richard Pohl was read out at the first performance, and is now printed in the full score; a good deal of it is the work of Princess Sayn-Wittgenstein.

It is shorter than the 'Faust' symphony, its two movements lasting about three-quarters of an hour, but it needs a slightly larger orchestra, including cor anglais, bass clarinet and two harps. The first movement, *Inferno* (which incidentally was copied very faithfully by Tchaikovsky in his *Francesca da Rimini*) begins with a musical setting of the words written over the gates of hell:

Per me si va nella città dolente;
Per me si va nell' eterno dolore;
Per me si va tra la perduta gente . . .
Lasciate ogni speranza, voi ch' entrate.

Through me is the way to the city of weeping;
Through me is the way to eternal torment;
Through me is the way among those that are lost . . .
Abandon hope, all ye that enter here.

The Italian words are actually written in the trombone parts, and the last line is made into a musical theme, Ex. 171, which plays an important part in the movement.

171

Then begins the musical representation of Dante's 'strange tongues, horrible cries, words of pain, tones of anger' which make up the whirlwind which ever rages in hell. A chromatic descending phrase, Ex. 172, is prominent at first, followed by hints of Ex. 173:

172

the tempo gradually increases till Ex. 173 appears in an *Allegro frenetico*, and finally reaches *Presto molto*, with wild rushing passages for the strings.

173

There is no abatement in the storm, Ex. 173 still remaining prominent, till a new theme, Ex. 174, also based on a descending phrase, bursts out on the full orchestra.

174

It is answered by Ex. 172, which gradually works up to a climax and the return of Ex. 171 on all the brass in the slow tempo of the beginning. The storm now subsides, and drum beats in the rhythm of Ex. 171 lead to the central section of the movement.

Like Tchaikovsky after him, Liszt here portrays the unhappy Paolo and Francesca, whose punishment for their illicit love is to be incessantly driven about by violent winds in the second circle of hell. He begins with an introductory section presaging Ex. 175, first on violins in 5/4 time, followed by a bass clarinet solo answered by clarinets in thirds; then after repeating this whole section he introduces a new theme, Ex. 175, on the cor anglais to Francesca's words:

Nessun maggior dolore
Che ricordarsi del tempo felice
Nella miseria.

There is no greater pain
Than to recall the happy days
In time of misery.

175

The resemblance of this theme to Ex. 174 is obvious; it is then repeated by bass clarinet and bassoon followed by flute and oboe in canon. A cello passage leads to an *Andante amoroso* section in 7/4 time; here Liszt portrays the lovers'

past happiness, and it is one of his most inspired passages. Its theme, Ex. 176, is again based on a descending scale.

176

After the whole section has been repeated with richer orchestration, Ex. 171 is heard sinisterly on the stopped horn and a harp cadenza heralds the return of the tempest. This begins with low horns and bassoons, and Ex. 173 soon appears, embellished with trills. Low clarinets and bassoons create an ironic atmosphere, and the storm soon reaches its height and continues on its course much as before. Finally Ex. 172 used as a ground bass works up to a last *fortissimo* restatement of Ex. 171.

The second movement, *Purgatorio*, begins with an introductory passage, marked *Tranquillo assai*, which seems to represent the scene where Dante, on leaving Hell behind, comes out into the light of the stars and sees the dawn rising like the 'sapphire of the orient'; it is a passage of wonderful beauty. After its repetition a semitone higher, we reach the portrayal of the trials which souls in Purgatory must endure in order to reach Heaven. A chorale-like theme on the clarinets and bassoons, Ex. 177(*a*), and its continuation on the strings, Ex. 177(*b*), indicate the yearning of these souls for ultimate happiness.

177

Then a fugue on a subject, Ex. 178, akin to Ex. 174 and 175, gradually builds up to an impassioned climax, ending in a *grandioso* section.

178

An interlude for woodwind over violin arpeggios leads to a return of the chorale theme, Ex. 177, in the same sad mood as before, but with its continuation, Ex. 177(*b*), the atmosphere begins to lighten, and soon the women's chorus enters with a delicate woodwind and string accompaniment to sing the Magnificat. In this final section Liszt uses diatonic harmonies with a modal flavour, and the general mood is joyful but tranquil, dying away at the end in ethereal chords. A second ending, which Liszt composed later at Princess Sayn-Wittgenstein's suggestion, brings the work to a *fortissimo* conclusion, but destroys the whole atmosphere of the final section, where Dante, having passed through Purgatory, gazes at the heights of Heaven above him and hears its music from afar.

Though the 'Dante' symphony is not the equal of the 'Faust' in sustained inspiration, its many fine passages certainly place it among Liszt's best works.

10

CÉSAR FRANCK

(1822–90)

John Manduell

*

'EMOTIONALLY strong but structurally weak.' We are all used to hearing such conveniently glib and easy verdicts on the large-scale works of composers such as Tchaikovsky and Liszt, however invalid they appear on closer examination. But in the case of César Franck, and in particular his only symphony, such a comment goes straight to the heart of the matter.

Franck was born in 1822, eleven years before Brahms and nineteen before Dvořák. Yet when in 1888 he completed his only symphony, Brahms had already written all his four symphonies and Dvořák's G major symphony was only a year off. Now Franck's training, like that of Brahms, was entirely founded in the classical school which was in turn reflected in his teaching, and yet when his turn came to write a symphony he was unable to build with the impressive architectural strength he so admired in others. But on the other hand, like Dvořák, he possessed the great virtue of a pronounced personal idiom, and it is the sheer strength of his music's character, with its heavy chromaticism, which so many have found so compelling.

Franck's Symphony in D minor was unusual for its time in having only three movements. The first alternates between *Lento* and *Allegro*, the second is a slow movement which embraces a scherzo and the finale is an *Allegro ma non troppo* but with the tempo often broken to admit of a backward glance at things heard in the earlier movements. This harking back is in accordance with the principles of cyclic form which Franck constantly favoured and did much to establish, and which aimed at securing a greater sense of unity by using the same ideas in more than one of the movements. In the finale of the symphony Franck will be found to use melodies from both the preceding movements.

But it is not only the principles according to which a composer works which interest us. There are the smaller details such as those particular fingerprints which immediately proclaim a prominent composer's authorship – such a fingerprint in this symphony is Franck's fondness for weak-beat accentuation, which two of the most important tunes in the work (Exs. 183 and 187) illustrate very clearly. And there are the larger considerations such as those of character, structure and orchestration which, of course, ultimately play such a large part in determining how highly one values any particular work.

Franck's orchestration has often been said to have an 'organist's thickness'. Certainly it tends to be rather solid and rarely shows any remarkable imagination, but it is always consistent and handled with a thorough assurance. As suggested, the construction of the symphony is not altogether satisfying. The lack of strong contrast in its melodic material, the absence of real organic growth in the first movement's development section and the deliberate recall in the finale of themes from earlier movements at whatever cost to the tempo structure; such are legitimate grounds for some dissatisfaction. But at the same time the reasons for the symphony's widespread and enduring popularity are not hard to find. It has its own very distinctive melodic appeal, it is patently shot with a burning sincerity and above all there is its stern but appealing character and the magnificent way in which, in typically Franckian manner, it rises gradually from its troubled opening to the triumphant sweep of the final pages.

The first movement opens with this rather brooding utterance by the lower strings:

179

on the last bar of which the violins answer with this lesser but subsequently fruitful phrase:

180

The violas and 'cellos underline the unrest with a quiet *tremolo* figure and then the opening (Ex. 179) is taken up by bass-clarinet, bassoons and horns and echoed a third higher by cor anglais, clarinets and bassoons. Now, after a *crescendo*, the music suddenly moves into a much quicker tempo, *Allegro*, as all the strings seize savagely on the opening theme, though its shape is altered at the third bar. This only lasts for twenty bars, although we might notice in passing a little phrase, typically chromatic in character, which Franck uses extensively:

181

The music is now pulled back once more to the opening slow tempo with Ex. 179 again quietly in the hands of the lower strings, now in F minor. This second slow section is closely patterned on the first until the *Allegro* is resumed. This, too, follows the previous pattern until the key changes to F major and a brief moment of more diatonic calm is reached:

182

Franck briefly expands this, making increased use of the dotted rhythm introduced at the fifth bar and then the music surges up in readiness for this tune, the one most people will probably hum or whistle first when they think of this symphony, the so-called 'faith motif', triumphantly proclaimed by the violins and upper woodwind supported by the trumpets below:

183

It is perhaps worth noticing the number of times the starting note of this well-known tune is heard. One of Franck's most marked characteristics is the way his themes either rock to and fro on a central note from which the tune never moves very far or else curl quickly back on themselves (e.g. Ex. 188). A Franck melody seldom contains any wide intervals; rather, it essentially uncoils.

Now follows the main development section. It is lucid enough not to present any difficulties in following but, as suggested, it does lack genuine growth and tends to advance somewhat by stops and starts, to proceed, in fact, rather than progress, until Franck pulls us suddenly back to the opening *Lento*, again in D minor. This time it is dominated by the brass, who throw the opening theme about in canon. Once more, however, it is not long before a quicker tempo is resumed, now in E flat minor. For a time the tension is relaxed while Ex. 181 returns in D major. Then the 'faith motif' takes over again as, with one or two quiet side glances, the music surges towards the last page. This is yet another return to the original slow tempo with the opening theme hammered out for the last time, again in canon but also with the notes twice as long as before – a perfect organist's conclusion, if you like.

The second movement is by common consent regarded as the most successful. It combines the functions of slow movement and scherzo with a happy ease and simplicity and this owes much to the fact that there is a more or less consistent basic pulse throughout, in contrast to both the outer movements.

After a short introduction of plucked chords on strings and harp, the main tune arrives on the cor anglais. This is the point at which the Parisian academicians spluttered with indignation at the first performance in 1889. No 'real' symphony would employ a cor anglais, protested the reactionaries! At all events, Franck has virtually ensured that we should accept the inclusion of the cor anglais by giving to it this very beautiful tune, which is admirably suited to the instrument:

184

This is then echoed by clarinet and horn, with the flutes joining in a few bars later. At the double bar which follows there is a change of key but, significantly, no change of tempo as the violins are given a long, rather sweet and typically Franckian tune to sing. Below them the 'cellos and double-basses have a rather broader line:

185

On this Franck now builds at some length. Then, as the texture thins once more, we come to the first hint of the scherzo, in the shape of a delicate figuration in triplets for the violins.

This scherzo section has a charming simplicity and nimbleness. Once the violins have their dancing triplets truly under way, the two clarinets introduce the main scherzo subject; again, the 'cellos have a broader supporting line:

186

Between the clarinets and 'cellos the violins continue their scurrying in triplets while horn chords complete the harmony.

The rest of the movement consists of a delightful sequence of play and interplay upon all this material, notably when the cor anglais returns with Ex. 184 to ride serenely over the still twittering violin triplets. Finally, as the triplets disappear, the music, after one last major climax, subsides quietly on a chord of B flat major.

The finale opens in radiant contrast to the rather introvert and heavily-scented atmosphere of the first movement. After six swashbuckling bars establishing the key of D major, the 'cellos backed by the two bassoons give us the main tune of the movement, perhaps the most genuinely joyous and uncomplicated tune Franck ever wrote:

187

Round it Franck builds an exhilarating climax. It is then succeeded by the second new idea of importance in this movement, heard on the trumpet supported by some of the other brass:

188

Like Ex. 187, Franck marks it *dolce cantabile*. But how different it is. We are immediately back to the half-cringing type of melody we so often find in Franck's music, and the mood of the symphony again becomes heavier and clouded. Nonetheless, such is the nature of Ex. 187 that each time it returns it is strong enough to dispel this mood so that in the end the overall impression is one of confidence and optimism.

Now the first of the cyclic cross-references appears with the cor anglais recalling its slow movement melody. The aim of this device, an increased sense of unity, is of course admirable if it succeeds easily and naturally. But if the procedure seems too obviously self-conscious, or if the natural flow of the music is broken unnecessarily in order that a theme from a previous movement can be accommodated, then it becomes in danger of defeating its purpose, as tends to happen here. Still, it is a good tune which most listeners will be glad to hear back again. It appears twice. The first is a brief occasion after which interest swings back to Ex. 187; but the second appearance is made the point of a big climax.

As this subsides, the violins quietly return to the 'faith motif'. This time the transition is an easy and natural one, and the tune's importance here is increased since it is made the centre of the ensuing climb towards the final climax, while 'cellos and double-basses introduce a steady but compelling *ostinato* in minims. Gradually the movement's initial momentum is fully restored and, with Ex. 187 brought back to crown the proceedings, the symphony ends, in D major, in an exultant blaze of triumph.

II

ANTON BRUCKNER

(1824 – 96)

Deryck Cooke

*

ANTON BRUCKNER was one of the great refashioners. There is not the slightest ground for the widespread view that his unorthodox structures are merely a spectacular mishandling of orthodox symphonic form. He knew well what he was doing, and that it was something unprecedented. He once said, with his characteristic simplicity: 'They want me to compose in a different way; I could, but I must not'. That he could is manifest in his early and 'orthodox' Overture in G minor, but when he turned to the symphony, he found himself forced along a new path. To express his own personal vision, he had to create a new type of symphony – vast, stark, and rough-hewn.* In consequence, he had to undertake one of the most difficult tasks in the history of music; and there are moments when even his warmest admirers feel that he has temporarily lost the way. But it is his own way that he has lost, not the traditional one. It is useless to try and evaluate Bruckner by yardsticks applicable to Beethoven and Brahms: we must assess his symphonies according to what they are intended to be. And if we do this, we find that Bruckner offers a unique *kind* of symphonic experience, not provided by any other symphonist. We may approach this experience through a brief consideration of Bruckner's unusual personality.

* The size of Bruckner's symphonies is entirely a matter of time-scale, and is not, as with Mahler, mirrored in the scoring. Bruckner began with a modest orchestra of double woodwind, four horns, two trumpets, three trombones, timpani, and strings; from No. 3 onwards, he used a third trumpet, from No. 4 a tuba, from No. 7 Wagner tubas (as in *The Ring*), and from No. 8 triple woodwind. In No. 8 alone he used a harp.

The son of a poor schoolmaster in the village of Ansfelden, in Upper Austria, he came from the most primitive stratum of European society – the Catholic peasantry. In Metternich's reactionary Austria, this class was unaffected by the growing liberalism and sophistication of European life in general, and its best stock had retained its original characteristics practically unchanged since feudal times – an earthy identification with the vast power of nature, a slow and massive natural strength of character, a genuine humility, and an unquestioning, childlike faith in God. Bruckner, by virtue of his musical gifts alone, rose from his 'humble station': he became a teacher and organist, first in a small town, and later at the monastery of St Florian. In 1855, at the age of thirty-one, he was an obscure provincial figure, who had acquired, with peasant-like tenacity, a formidable knowledge of the almost obsolete 'strict' methods of composition and a large number of examination diplomas; he had also composed a considerable amount of music, mainly liturgical and almost entirely academic.

The following year, however, his appointment as organist of Linz Cathedral brought a completely unforeseen turning-point. In this bustling town, he became more familiar with the music of Beethoven and – even more important – he heard some Wagner for the first time. The tremendous latter experience acted as a catalyst which brought about the crystallisation of his latent genius, though this did not result in an adoption of Wagner's style or the medium of the music-drama. He grafted elements of Wagnerian romanticism on to the solid technique he had already acquired – the pure harmonic and contrapuntal style stemming from the sixteenth century. In this way, during 1864-8, he found his individual voice, and composed his first mature masterpieces: the Masses in D minor, E minor, and F minor. He was already over forty.

At this time too, he composed his first mature symphony – No. 1 in C minor; and having gained a professorship at the Vienna Conservatoire, this devout church-composer and superb organist settled in Vienna and became almost exclusively a symphonist. In the fashionable Austrian capital he cut an anomalous figure, retaining the habits of mind, the manners, and even the clothes of the simple rustic; with his naïve Catholic faith, he was utterly unaffected by the social and political currents of his time, and by the seething intellectual and emotional unrest of the romantic movement. He was a survival from the past – a primitive, in the exact and lofty meaning of that word. In composing his nine symphonies, he was not the contemporary type of 'artist', refracting the spirit of the age through a flamboyant individual personality; he was a highly skilled craftsman, erecting to the glory of God huge cathedrals in sound, the intense inspiration of which was not so much a purely personal experience as a drawing on the simple but profound wisdom and spirituality of his ancestors.

The startling novelty of these works awakened acute critical hostility, the more so because Bruckner was dragged into the Brahms-Wagner feud, to his own bewilderment. He was set up, for praise or blame, as 'the Wagnerian symphonist', in spite of the fact that Wagner's artistic theories meant nothing to him, and that his music was entirely un-Wagnerian in spirit. Not until 1884, his sixtieth year – when the first performance of the Seventh symphony, in Leipzig, was received with unanimous enthusiasm – did he begin to be recognized as an individual genius. In his last years he was revered by the Austrians as 'the master'; after his death, in 1896, this acclamation was echoed in Germany and Holland; but the rest of Europe has been reluctant to join in, except for a growing body of admirers in Great Britain.

Part of the reason, undoubtedly, has been the vexatious 'Bruckner problem' – the confusion over the various versions which exist of most of his symphonies. Owing to the initial difficulty he experienced in balancing the proportions of the new kind of structure he was creating, he produced revised versions of some of the earlier symphonies during the period of his growing mastery. Unluckily, in his extreme humility, he let himself be persuaded by clever colleagues and pupils that the resistance to his symphonies might be overcome if he made extensive cuts in the form and changes in the orchestration; and worse, his advisors took his willingness as *carte blanche* to produce mangled versions of their own. The first editions of Bruckner's symphonies, published between 1878 and 1903, are nearly all of these kinds.

Since these scores could not claim to represent Bruckner's intentions, the Bruckner Society was founded in 1929 to publish his own revisions which, he decreed in his will, should be left for later times; between 1934 and 1944, all nine symphonies were issued, except No. 3, mostly edited by Robert Haas. Thus matters were set right, it seemed, and there was no longer any problem. But unfortunately, owing to dissension in the Bruckner Society, Haas was replaced, and since 1950 a new series of 'original versions' has been issued – all nine symphonies except No. 2 this time – mainly edited by Leopold Nowak. In general, Haas eliminated all alterations which he believed had been wished on to Bruckner, but Nowak retained some of them, regarding them as Bruckner's own final intentions.

Whether Haas or Nowak best represents Bruckner is a matter for debate; however, the problem should not be exaggerated.* The important point is that, from now on, the first editions are utterly discredited (apart from Nos. 6 and 7, which contain only minor discrepancies); the worst offenders are Nos. 5 and 9, arrant falsifications made without Bruckner's own cooperation.

The situation has been bedevilled, however, by the vagaries of conductors. Not only do some of them still use the discredited early editions, but they are prone to make cuts of their own; and worst of all, they often play fast and loose with the tempo as they never would with any other composer. In general, German conductors, impelled by exaggerated reverence for the spiritual element in Bruckner, adopt inordinately slow tempi which deprive the music of all vital impulse, and often slow down still further for the lyrical elements, as well as making heavy rallentandos and accelerandos at focal points, which are not specified in the

* The present writer adheres to Haas.

authentic versions. The reader is warned that when he hears a Bruckner symphony, he should be wary, and consider carefully whether the conductor is getting enough momentum into the music to make it register properly. Admittedly, Bruckner's monumental style demands broad tempi; but if his music should never be hurried, it should also never be dragged, and each movement should be taken in a single basic tempo, unless marked otherwise, without abrupt changes of gear.

Having groped through the fog which obscures Bruckner the symphonist, we can now consider the symphonies themselves, and try to discover what it is that makes them so unique. The essence of them is that they express the most fundamental human impulses, unalloyed by civilized conditioning, with an extraordinary purity and grandeur of expression; and that they are on a monumental scale which, despite many internal subtleties and complexities, has a shattering simplicity of outline. These fundamental characteristics are peculiar to Bruckner, though the size and the grandeur owe something to Beethoven and Wagner. Beethoven's Ninth left an indelible impression on him, and partly conditioned the layout and character-types of his movements – the far-ranging first movement, the big Adagio, the bludgeoning Scherzo, the huge cumulative Finale. Wagner inspired him no less, giving him an idea how to build up extended harmonic periods, how to use the brass for weight of utterance, and how to get at the root of an emotion with a deeply *espressivo* type of string writing. (These Wagnerian technical features, incidentally, are the three things – the only three – which Bruckner shares with Mahler, with whom he has for so long been pointlessly coupled.)

A good starting-point for our investigation of Bruckner's special quality is the famous *bon mot* that he composed, not nine symphonies, but one symphony nine times. This, like all such witticisms, contains a grain of truth: the fact is that with Bruckner, certain types of material tend to recur from symphony to symphony more commonly than with other composers – a feature which, incidentally, confers on his style a remarkable homogeneity. But the important point is that these things are indeed *types*: each individual example is quite different from the others, and the context in which any given type occurs is different from symphony to symphony, so that its significance is never twice the same. Moreover, there is much else in Bruckner besides these recurrent types; each of his symphonies has its own marked individuality.

The most convenient way to see Bruckner in the round is to examine examples of his four movement-types, and at the same time to indicate the kinds of material he uses as they occur. I have chosen, not a single symphony, but four movements from different symphonies; this will enable us to see Bruckner at his most characteristic, and also give us a wider view of this still little-known *oeuvre*.*

The first movement of No. 3 opens with four introductory bars which build up a multiple *ostinato* on the tonic chord – the key-chord of D minor.

189

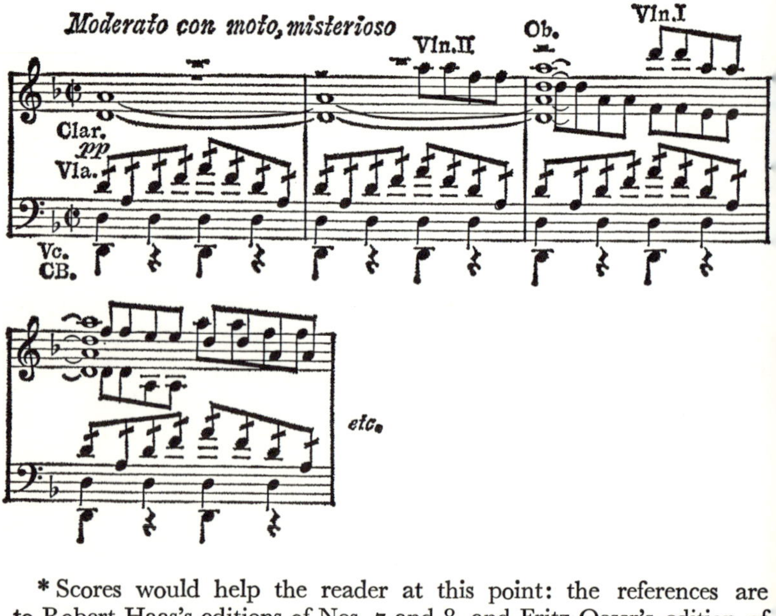

* Scores would help the reader at this point: the references are to Robert Haas's editions of Nos. 7 and 8, and Fritz Oeser's edition of No. 3 (all published by the Bruckner Society).

This is Bruckner's characteristic way of beginning: the main theme does not appear straight away, as nearly always with other symphonists. There is another *bon mot* to the effect that Bruckner could not get a symphony under way without a string *tremolando*; and again the shaft contains a partial truth. The mysterious opening of Beethoven's Ninth held a tremendous fascination for him:* five of his nine symphonies do begin with a string *tremolando* – and so do several of his later movements. But the category really needs extending to embrace the whole idea of a background accompaniment, including a rhythmic figure or an *ostinato* (cf. Mozart's Fortieth symphony); and then it is true that all but one of Bruckner's symphonies and most of his later movements begin in this way. It may be noticed that the opening of No. 3 (Ex. 189) is not a *tremolando* but, as stated above, a multiple *ostinato*.

It hardly needs saying that Bruckner's use of this general procedure was not due to inability to begin otherwise: in one of his greatest symphonies – No. 5 – not one of the four movements opens like this. But it is part and parcel of his personal vision that his music tends to emerge gradually out of silence, as a neutral background sonority evoking the mystery of creation itself. Then, against this background, the act of creation takes place with the presentation of the main theme.

While the *ostinato* of Ex. 189 continues, the main theme enters in bar 5 on solo trumpet, followed by pendants on woodwind and horn.

190

* Strictly speaking, the Beethoven opening is not a tremolando, but a defined rhythmic reiteration in sextuplets. The effect is much the same, however.

This too is entirely characteristic. It is what the Germans call an *Urthema*, or 'primordial theme' – a stark motif on the tonic harmony, revolving around the octave, the fifth, or the complete triad (cf. the opening themes of Beethoven's Ninth and Wagner's *Rheingold*). Five of Bruckner's symphonies begin with a theme of this kind against a background, but there are crucial differences. It may be used as a preludial motive, for example (No. 9); and other symphonies open quite differently. No. 1 begins with a march-theme over a rhythmic pedal, No. 5 with long-drawn counterpoints over a *pizzicato* bass, No. 8 with separated bass motives against *tremolando* outside the main key.

However, the opening of No. 3 is typical in its majestic two-in-the-bar pulse and the bold simplicity of its continuous tonic-chord *ostinato* and stark unadorned *Urthema*. To appreciate this naked emphasis on the pure fundamentals of musical language demands an entirely fresh approach: it is the essence of Bruckner's startling originality, in the way it digs right down into the roots of things.

Ex. 190 takes us to bar 17, but the tonality has still not budged. Where the horn enters, in Ex. 190, the chord changes to B flat, and two bars later to the dominant chord of A, but the keynote D persists in the bass, and will do so until bar 31. This is the typical Bruckner procedure – the basing of the whole long opening paragraph on a tonic pedal (cf. the opening of Wagner's *Rheingold* and *Walküre*). It gives his symphonies their massive, solid foundations and sets their vast scale. For the impatient, no doubt, the prospect of waiting so long for the initial tonality to change is a daunting one; but then, Bruckner's far-flung splendour is beyond the understanding of the impatient.

The way this large-scale procedure actually works offers another typical feature, which can occur at many different places in a Bruckner symphony. A high pitch of tension is worked up by persistent repetition of a short rhythmic phrase in progressive diminution, with a full orchestral *crescendo*. At the end of Ex. 190, phrase A begins overlapping on woodwind and horn; then the shorter two-note phrase B is worked up, over the persisting dominant harmony with tonic pedal, in the following way. (Note the repeat marks, used here for reasons of space.)

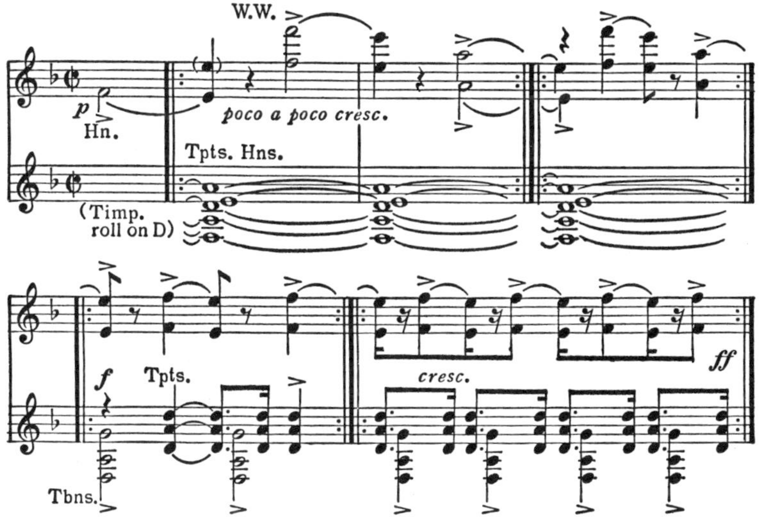

191

Only the wind and brass outline is shown: the strings are also hammering out their original *ostinato* figures. This is the characteristic Bruckner 'blaze-up', which contributes much to the sense of primeval, elemental power in the symphonies; it is greatly enhanced by the highly individual, entirely un-Wagnerian orchestration, in which each group of instruments stands out vividly with its own particular rhythmic figure. The sophisticated may find the procedure stupefyingly obvious, but if one yields to it without preconceptions, its impact is electrifying; and it could not be achieved, of course, by any less simple means.

Ex. 191 floods over into the climax – a new idea of shattering strength.

192 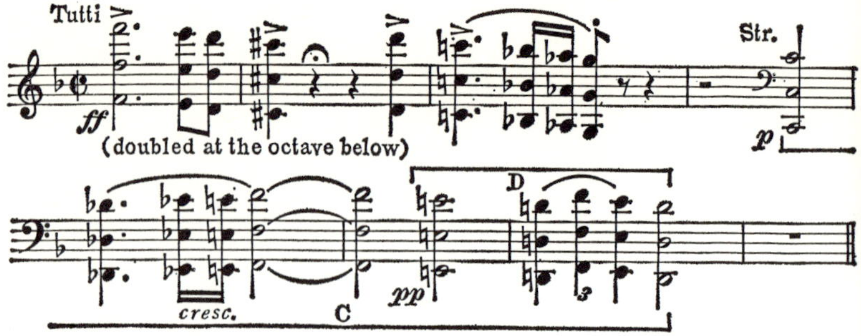

This is a further thematic type – the tremendously weighty theme for the full orchestra in unison, with pauses and rests between the phrases. It also can occur elsewhere; in No. 7, for instance, it appears as the third idea of the finale. The type clearly derives from the great unison outburst which is the main theme of Beethoven's Ninth, but Bruckner made the conception all his own.

The unison theme is restated, with full harmony; then, as with the first theme, it proceeds by repeating its later phrases (C and D of Ex. 192), but this time as dying echoes – a dissipation, not a gathering of tension. The contrast here is enormous, between the uninterrupted momentum and growing intensity of the opening, and the discontinuous phrases and increasing quietude of the new idea.

By bar 58, we have still not abandoned the tonic key. This whole extended double paragraph represents the typical first section of a Bruckner opening movement: it corresponds to the 'first subject' of sonata form, but it is different in that it presents, not one theme, but two completely antithetical ones. A short link passage – a violent outburst based on D (bars 59–67) – leads to a restatement of the whole first section. This is shortened considerably (34 bars as against 58), and moves through other keys, rather like the 'transition' section of sonata form. But its effect is utterly different: it does not now work up tension to prepare the 'second subject', but merely provides a counterbalancing restatement of the first section. *Ostinato*, *Urthema* and 'blaze-up' are now based on a dominant pedal, A, and the unison theme enters this time in B flat, its echoes dying away in an indeterminate tonality.

As we reach the second section, Bruckner's unique formal method becomes clear. He presents his materials in separate blocks, cheek by jowl, often without any dovetailing of one into the other, as is the normal practice. The indeterminate chord reached by the dying echoes of the unison theme is left hanging in the air, as the second section begins in a new key; the lack of any smooth transition or dynamic follow-through is intentional. Whereas, with other composers, the new key is usually established and clinched with a forceful modulation aided by rhythmic tension, Bruckner introduces his new key of F major statically, as a kind of subsidiary perspective of the basic D minor. It is this kind of procedure that gives his symphonies their sense of equanimity and deep peace at the heart of turmoil. Following on the indeterminate chord, the second section accepts this harmony as implying F major (note how it picks up the unison theme's triplet rhythm).

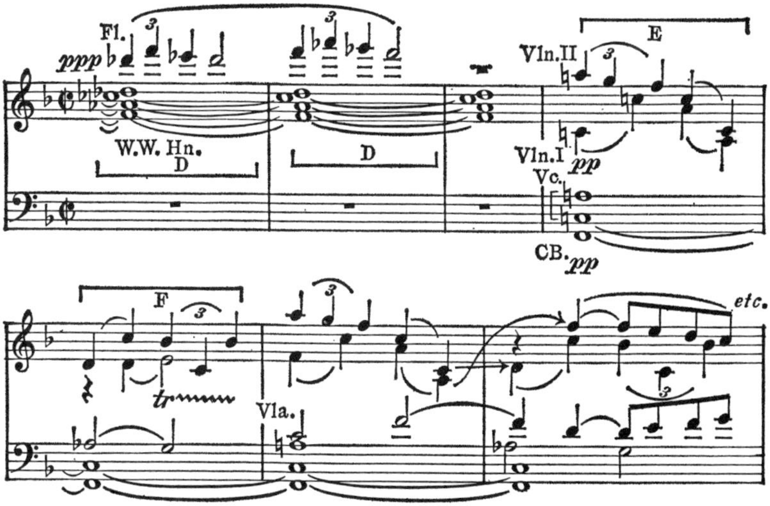

193

This second section introduces another thematic type – the lyrical, flowing melody, which may be gentle, graceful, tender, warm, impassioned, or all of these at once. Bruckner called this section the *Gesangsperiode* (the 'song period'). It often, as here, takes the form of two or more melodies woven together in simple counterpoint; although the violas clearly have 'the real theme', the first and second violins also have attractive melodic lines of their own. The result is another enormous contrast – between the stark, insistent, short-spanned themes of the first section and the pliable, gentle, long-drawn melodies of the second. Moreover, whereas the first section was firmly rooted in its basic key, the second section, after beginning on a (new) tonic pedal of twelve bars, begins modulating extensively as it flowers and develops: during its seventy bars it moves through G flat, E, A flat and D flat, to return to F, whence it set out. This procedure is only possible because of the massive tonic foundation of the first section. The end of this second section is a 'blaze-up' on phrase E, followed by another on phrase F, to initiate a new idea, in the minor mode of the new tonic F.

194

This is the characteristic third section – for Bruckner's expositions usually consist of three clearly defined sections, as opposed to the normal two of sonata form. It returns to stark unison material, to counterbalance the first section, but gives it much more drive and sustained rhythmic impetus. The way the strings (*marcato*) and the wind and brass (*legato*) have different figurations of the same basic line is the essence of Bruckner's pounding third-section unison textures, creating a curiously evocative bareness of sonority. It will be noticed that the theme here continues the two-plus-three rhythm of D in the second section, which is one of Bruckner's fingerprints.

The third section too is typically long-spanned (bars 171–256), and modulates extensively, but it has two unusual features. First, it encloses a further new idea – a chorale-theme on the brass against the basic unison motive (bars 201 ff.); secondly, it culminates in a return of the *Urthema*, in E major (bars 211–19). Then it follows Bruckner's normal course of dying away in the key of the second section, F major. There has been a wonderful sense of homogeneity about the vast but simple D minor/F major progression of the exposition.

A double bar here seems to indicate that the development is about to begin. But although Bruckner's first movements do fall into three main stages, their connexion with the exposition, development and recapitulation of sonata form is deceptive and misleading. The problem of the late nineteenth-century symphony was that the exposition already contained so much development that the development section itself ceased to have any genuine *raison d'être*. Bruckner solved this problem by making his last two stages refashionings of the exposition from different points of view. Both the 'development' and 'recapitulation' are usually 'developed restatements' which combine with the exposition to form one single, continuous, movement-long process.

To analyse the rest of the movement is beyond the scope of this essay, but the following main elements may be pointed out. After the double bar, there is first the characteristic sense of deep peace, with echoing silences (bars 257–67); next the typical awakening to life – a development of the first section's *ostinato*-and-*Urthema* and unison theme, in F minor and G minor (bars 268–98); then a new development of the unison theme against *pizzicato* strings, in D minor and C minor (bars 298–340), which culminates in a 'blaze-up' on the dominant, A. And here occurs the deception. The *Urthema* breaks out in the tonic in shattering full orchestral unison, without its *ostinato*, and we expect this to be the recapitulation; but actually, what has gone before is only an episode – we have now reached the beginning of the second stage proper, the first of the two 'developed restatements' of the exposition. This presents the exposition's first section from an entirely new point of view (bars 341–402), and also the second section, briefly (bars 403–

28); the third section is omitted. Then the final stage begins, with the *ostinato*-and-*Urthema* quietly in the tonic as at first: more faithfully than is usual with Bruckner, this is an orthodox classical recapitulation, restating the exposition almost exactly, but with contractions, and with all the materials now in the tonic key (bars 429–588). Bruckner was to handle the continuous three-stage process more originally in his later symphonies, and also to initiate his second stage more masterfully than in the case of the episode mentioned above, which is not particularly convincing. The movement shows Bruckner at his most characteristic, not at his greatest; yet its impact is hardly less than those of his great first movements, owing to his peculiar capacity for getting across what he has to say in spite of inadequacies of form.

The ending is magnificent, however. As so often with Bruckner, it is a twofold coda. There is, first, a development of the *ostinato*-and-*Urthema* over a dominant bass, which culminates in a 'blaze-up', typically unresolved, followed by a silence and a hushed reference to the unison theme (bars 589–625). Then, after a further silence, this dominant passage is answered by the characteristic Bruckner peroration, which makes the whole coda one gigantic dominant-tonic cadence. The final pages are a continuous twenty-four-bar *fortissimo* tonic chord: the strings have the *ostinato*, while the brass give out various versions of the *Urthema* (direct and inverted), in overlapping rhythms; the end is a last crowning 'blaze-up', with a typically abrupt break-off and echoing silence in the final bar's rest (Ex. 195 – note the final bare fifth harmony).

195

The effect of the extended tonic pedal at both the beginning and ending of the movement is that of two broad, solid bases on which the whole edifice rests; and this plays a large part in the impressive simplicity of outline mentioned above.

Usually, the opening movement is followed by the slow movement and then the scherzo, though in the last two symphonies Bruckner reversed the order, as in Beethoven's Ninth. The slow movement is nearly always a long-drawn, deeply-felt *Adagio,* based on two alternating sections; normally the alternation occurs twice, as in the *Adagio* of Beethoven's Ninth, and the final appearance of the opening section rises to an overwhelming climax, followed by an extended coda. The undeniable sublimity of Bruckner's slow movements has made them the most widely admired aspect of his art; the greatest are undoubtedly those of the last three symphonies, for which he added a quartet of Wagner tubas.

Perhaps the most characteristic is that of No. 7. The opening section is typical, a long-spanned paragraph comprising several distinct ideas, following one another logically in one glorious flow of thirty-six bars. Ex. 196 shows the first two: the tragic opening phrase in C sharp minor, for rich lower strings backed by the majestic tubas, and the contrasting answer, boldly confident, for strings alone, joined by the trombones.

196

This passage shows the extraordinary purity of Bruckner's handling of romantic harmony, deriving from his expertise in the old 'strict' methods. The calm third idea (bars 9–12, not quoted here) reveals his tonal subtlety: it sharply contradicts the subdominant tendency of Ex. 196 (towards F sharp minor) with a bright preoccupation with the dominant, G sharp major (A flat) and dominant's dominant, D sharp major (E flat). The fourth idea (Ex. 197) exemplifies Bruckner's powerful handling of the tricky business of sequence, owing to his mastery of tonal perspective: four repetitions of a one-bar phrase follow the most unexpected tonal progression, swinging back from D sharp to F sharp, and thence, sadly, to D major and B minor (the double-subdominant on the far side of the home key from D sharp major).

197

Two more ideas follow a hopeful one in the relative major, E (bars 19–22) and an aspiring one in the dominant again, G sharp (bars 23–32); but the latter rises to an anguished climax-chord, and then a magnificently dark passage for low tubas and solo horn (bars 33–36) leads into the second section – in the subdominant, F sharp, which was the first section's original tendency. This second section, in contrast to the first, consists of one single, flowing, heaven-sent melody.

198

Actually, the melody appears quite simply as the lower line of Ex. 198, doubled at the octave above (first and second violins); the small notes are a subsequent development, to be mentioned later. Its full statement (bars 37–61) occupies most of the second section, the rest being a brief coda (bars 62–76), which leads back to the first restatement of the opening section.

In fact, the 'restatement' is nothing of the kind, nor even a variation, but a development of the two ideas of Ex. 196: first of A (bars 77–104), then of the little three-note figure which ends B (bars 105–14), and finally of the whole of B (bars 114–32). The latter reaches a G major climax (bars 127–30), which foreshadows the eventual climax of the whole movement; it dies down for the restatement of the second section. Again in contrast, this is mainly an exact repetition, in briefer form (bars 133–56). The key is now the dominant, G sharp (A flat) – on the opposite side of the main key from the subdominant, F sharp, in which it first appeared; and the melody is joined by the counter-melody given in small notes in Ex. 198, so that the whole takes on the contrapuntal-melodic effect of the characteristic first-movement *Gesangsperiode*.

The culmination of the movement follows – the final restatement of the opening section – which is mainly development again. There are two typical procedures here: first, A is surmounted by a repeated slow violin figuration, which imperceptibly generates great tension; secondly, as this continues, the three chorale-like chords which begin phrase B (a Bruckner fingerprint) are developed by the brass through a remarkable extended sequence, following the most original tonal progressions. Ex. 199 gives a glimpse of this vast process.

199

The outcome is a mountainous *crescendo* and an incandescent C major climax (bars 177–82), in which the trumpets blaze out the little three-note figure of B. Then the splendour dies away; the tubas lead back mournfully to the tragic main key (bars 184–93), for an ethereally sorrowful coda based on Ex. 197, which has not been heard since its first appearance. The ending, as with practically every Bruckner movement, is a return to the opening idea (A): the first phrase is repeated several times, peacefully now, in C sharp major, over a 12-bar tonic pedal.

The typical Bruckner scherzo follows tradition in being in three sections, but is nevertheless of a kind all its own.

It combines rhythmic impetus with a great weight; and apart from the traditional dance-form type of the earlier symphonies, and unusual examples like those in Nos. 6 and 9, it develops a single short idea, which rises to a characteristic 'blaze-up' to round off both statement and restatement. The scherzo of No. 8 (which is actually the second movement of that symphony, as mentioned above) may be examined here, being the type *par excellence*.

Its short basic idea, of the *Urthema* type, has a *tremolando ostinato* for background.

200

Thus the idea is twofold – the *tremolando* figure A, and the *Urthema* figure B. After this four-bar statement, in C minor, it is subjected to a single non-stop working-up process right to the double bar. First, the music plunges straight into the surprising key of G flat to develop A through a big *crescendo* towards E flat minor, thus bringing the relative major of E flat itself in sight (bars 7–24); then B takes over quietly, and is developed through another swifter *crescendo* (bars 25–32) which switches suddenly into the even more surprising key of A major. From here to the double bar, the process is one of insistent repetition of the twofold idea, through a final tremendous *crescendo*, towards a climax in the orthodox tonality of E flat, the relative major. This passage is so fantastically original and so utterly characteristic that it deserves quoting in full. (The number of times each bar is repeated is indicated by a number in brackets.)

201

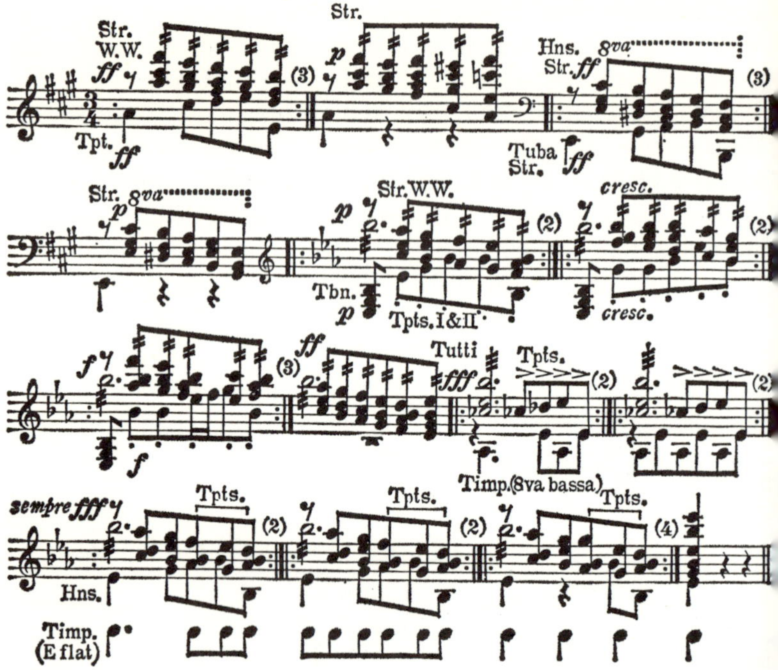

One characteristic subtlety should be noted here: when the music slips into E flat (in Ex. 201), there is a definite expectation of further modulations to come, but in fact there are none: the music has already reached its tonal goal almost unnoticed, and the remainder of the whole rhythmic period expends itself in the 'blaze-up', break-off, and echoing silence.

The orthodox central section follows – further development of A and B, of a completely different kind, hushed and mysterious in mood (bars 65–134). It arrives on a tense dominant pedal (horn, bars 123–34), which initiates the traditional restatement. And this offers another remarkable subtlety: it is exactly the same as the opening section, with the single exception that the move from A major to E major in Ex. 201 is replaced by a move from A major to *D flat major*, so that the ensuing drop of a semitone slips the music this time into the tonic C major for the final 'blaze-up'.

Bruckner's trio-sections are normally slow and quiet; in

the earlier symphonies, he used the traditional Austrian *Ländler* type, but from No. 5 onwards he created something new for each symphony. The example in No. 8 is a lovely flowing theme over a hushed march-like tread of the Schubertian type, on *pizzicato* strings. The beginning of the melody is given in Ex. 202; and anyone who imagines that Bruckner is a naïve composer should try his hand at harmonizing it (in A flat), and compare his attempt with Bruckner's intensely expressive but wonderfully pure progression.

202

Bruckner's finales display a greater variety between them than is the case with the preceding movements – so great that none can be regarded as a 'type'. Some are built on a similar pattern to the first movement, on a rather larger scale, but the finale of No. 7, for instance, is much more compressed; while the great finale of No. 5 is unique – a vast sonata structure in which the first section is a fugue on an octave-type theme, the second a *Gesangsperiode*, the third a fugue on a chorale theme, and the culmination a double fugue and final chorale-and-fugue apotheosis.

However, we may examine the finale of No. 3, returning to that work because with Bruckner there is a close relationship between first and last movements which needs to be considered. Since its form is similar to that of the first movement, it need not be analysed here, but we may glance through it – and encounter in the process three further thematic types.

As is often the case, the opening section differs from that of the first movement in being more urgent, and not having a tonic pedal for basis. It presents, against a characteristic *ostinato* (Ex. 203(*a*)), our first new thematic type – the great striding motif based on rising and falling ninths and/or octaves (Ex. 203(*b*)).

203 

An early example of this type of theme may be found in Wagner's *Faust* overture, but Bruckner was probably influenced by the Austrian classical tradition, especially Mozart (cf. the opening of that composer's 'Haffner' symphony).

Following a typical pause, the second section appears – a *Gesangsperiode*, but of a most unusual kind (and in a highly unorthodox key!). Against a lilting polka, played by the strings, the brass intone a solemn chorale (our second new thematic type, which has already appeared – unquoted – in the first movement).

204 

The third section proves to be a pounding unison idea, as in the first movement.

In the first 'developed restatement', the chorale melody of the second section (Ex. 204), now in the minor, loses its accompanying polka, and is merely supported by its harmonies; the result is our last thematic type – the characteristic *legato* string melody in an inner part, gently melancholy, threading its way through a quiet tread of detached or *pizzicato* string chords.

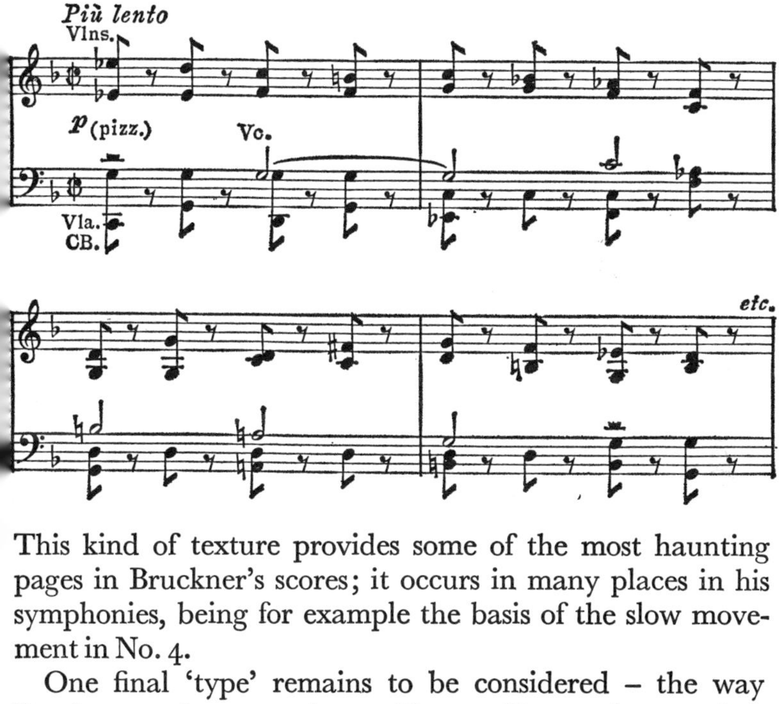

205

This kind of texture provides some of the most haunting pages in Bruckner's scores; it occurs in many places in his symphonies, being for example the basis of the slow movement in No. 4.

One final 'type' remains to be considered – the way Bruckner ends a symphony. No. 3 offers a characteristic example – a coda in the tonic major, devoted to a simultaneous brass presentation of the symphony's original *Urthema* (Ex. 190), in direct and inverted forms, as broadly as possible, against a blazing orchestral *ostinato* – the whole making up a prolonged 42-bar tonic major chord (with eight intermediate bars of subdominant). The effect is naturally to counterbalance the opening tonic pedal of the whole work, and to expand on the culminating tonic harmony that ended the first movement (Ex. 195).

This might seem the most naïve kind of cyclic procedure – throwing in the symphony's opening theme at the end as a sort of *deus ex machina*. But it is nothing of the kind. Bruckner's style is so homogeneous that the recurrence of earlier themes seems perfectly natural; and in any case, it is motivated here by a subtle relationship between the original *Urthema* and the opening theme of the finale – *both begin with the same decisive four-note rhythm* (cf. Ex. 190 and Ex. 203(*b*)). This is made clear by Bruckner's brief but powerful reference to the finale's opening theme as an introduction to the coda (bars 589 ff.). It should be stressed that the restatement of the original *Urthema* is not always at the very end, like this; in Nos. 5 and 8 it fulfils a much more subtle function.

So much for the 'characteristic' Bruckner symphony. It only remains to say that there is, of course, no such thing. The music-lover with the patience to explore Bruckner's complete *oeuvre* will discover as fascinating a variety as anyone could wish; and he will be rewarded, it may not be superfluous to add, by an experience of a uniquely inspiring and uplifting kind.

12

ALEXANDER BORODIN

(1833–87)

John Manduell

*

IF we paraphrase the title of Molière's comedy we might briefly describe Borodin as '*le compositeur malgré lui*'. Certainly throughout his life his chief preoccupation was with medicine and in particular chemistry, of which he was a widely-respected professor. He was therefore a spare-time composer and this was both his strength and his weakness; his strength in that it enabled him to remain relatively uninfluenced by the music of others and his weakness in the sense that he might obviously have achieved so much more had he had the time to do so. In fact he left only four purely orchestral works, including the three symphonies.

His lack of formal training did not prevent him from becoming an accomplished craftsman. He was an amateur composer only in the true sense of the word as indeed, to a greater or lesser degree, were his close musical friends who with Borodin made up the famous group 'the Five', Cui, Balakirev, Rimsky-Korsakov and Mussorgsky. And if he did not reject the many offers of help he received from the more experienced Glazunov and Rimsky-Korsakov in particular this does not mean that his touch was not sure. It almost always was, and his music has that kind of strength that when the unexpected occurs it appears inevitable. He displayed a fine mastery of form, and in fact his ability to bestride a large canvas with such consistent ease when leading such a disjointed musical life is remarkable.

His strong sense of construction was backed by a pronounced personality in which certain qualities predominate, such as his natural command of the epic and the tender and his apparently effortless ability to blend these two extremes in almost one and the same breath, his Euro-Asian melodic style stemming, one assumes, from his Russian and Caucasian ancestry, his rugged sense of rhythm and his forthright but by no means unimaginative orchestration with its dramatic contrasts owing much, one senses, to his love of the theatre. Such are the elements which, in a happy fusion, have ensured that his two completed symphonies, and in particular the Second, should enjoy such enduring popularity.

Borodin's First symphony (in E flat) occupied him from 1862 to 1867 and was first performed under Balakirev in 1869. It is hardly surprising that it should have taken a spare-time composer five years to write his First symphony. What is more remarkable is that this, his first work for orchestra, should have been accomplished with such assured control of the medium.

At the second bar of the slow introduction, bassoons, 'cellos and double-basses embark on the main melody of the first movement:

206

As the movement progresses Borodin makes great use of the upward figure at the third bar and also of the second bar, primarily for rhythmic purposes. In fact one of the most immediately striking aspects of his craftsmanship is the way in which he repeatedly breaks a melody down into its individual sections and uses these fragments independently. This is well illustrated by the start of the *Allegro* which follows the slow introduction where after two bars the timpani embark on a steady reiteration of bar 2 whilst two bars later violas and violins begin tossing bar 3 about among themselves. The rhythmic drive is maintained until, at an abrupt *crescendo* to *fortissimo*, Borodin introduces the threefold group of second subject ideas. The first is a heavily accented rising figure in crotchets, the second is the lyrical and chromatic passage for strings which follows and the third is a brief but very characteristic two-bar phrase first heard on the oboe immediately after this passage for the strings. A full restatement of Ex. 206 follows at the next climax and then Borodin launches the development, again significantly with a steady repetition of the second bar of this melody, this time by the clarinets. The development is a finely-controlled marshalling of all the ingredients we have heard, and again one applauds the remarkable sureness of touch of a chemistry professor writing his first orchestral work. The recapitulation when it comes is so full of natural invention as to be virtually a continuation of the development, and this then leads to the coda in which, by the simple means of quietly bringing back the opening melody but now with the notes doubled in length, Borodin achieves an affectingly tranquil echo of the preceding restless agitation as the movement ends simply with a woodwind chord of E flat major.

As mentioned, Borodin was relatively uninfluenced by the music of others. But he was at this time particularly fond of the music of Mendelssohn, and the scherzo (again in E flat major) which follows seems to some extent indebted to Mendelssohn. It is light and airy and it never loses its sparkle, but it is hardly characteristic of Borodin. Its chief ingredients are all very simple: a *staccato* ascending figure in quavers for the strings, a complementary descending scale for flute again in *staccato* quavers and a simple four-note descending figure which Borodin uses extensively.

The trio in B major in the middle of the movement is, however, very different. Here one can glimpse the Borodin of *Prince Igor*, and this whole section has an originality of melody, harmony, orchestration and irregular metre which raises its interest far above that of the flanking *prestissimo* sections. Here is the main melody of the trio as heard at the beginning on the oboe:

207

We have already remarked on the help, if this is always the right word, which friends such as Glazunov and Rimsky-Korsakov gave Borodin, particularly over matters of scoring. The third movement is a case in point. The main tune was originally scored for cor anglais. As garnished for our delectation in the edition by Glazunov and Rimsky-Korsakov, it is presented on the 'cellos. How much more effectively in keeping with the tune's Russian folksong character the cor anglais would have been becomes apparent when later in the movement, as the scoring now stands, it is permitted to lend its own distinctive quality to this melody.

This very beautiful movement, in D major, is perfectly simple to follow. It consists of easy natural mutations of this main tune with much ornamentation by means of the demisemi-quaver turn in the sixth bar. Here is the melody as sung by the 'cellos at the beginning of the movement:

208
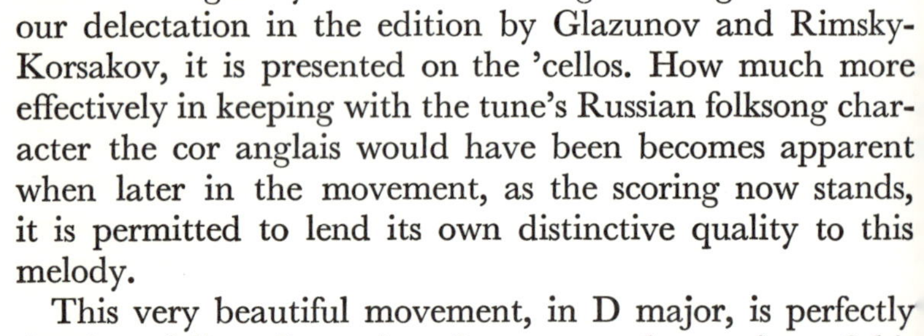

The finale, again in E flat major, is an invigorating well-compounded movement, straightforward in construction but with a commanding strength and impetus as it sweeps along to a fine storming finish. If it is not altogether representative of Borodin at his most individual, this is perhaps due principally to two factors: its use of many harmonic and sequential clichés such as one can find in the works of any number of nineteenth-century composers, and the rather square and not particularly distinctive first subject which dominates the movement from the very first bar. The second subject, however, with its deliberate weak-beat accentuation, carries more of the composer's own stamp. It first appears on clarinets, bassoons and strings in octaves immediately after a descending sequence of chords at the climax to the statement of the first subject, and it becomes increasingly important as the movement progresses:

209

The composition of his Second symphony, in B minor, was again, in view of Borodin's other preoccupations, a protracted affair, lasting from 1869 to 1876. But the result is undoubtedly a highly individual masterpiece.

In the first movement, inspired we are told by the image of a gathering of eleventh-century warriors, Borodin once again introduces us without any preamble to the first subject. It is really two short phrases which begin:

210

These short phrases are submitted to a wide range of varied treatment, often very simple and always effective. The second bar, for instance, appears later in strongly accented crotchets and then in the coda in martial minims, whilst on the other hand it can, at another point, be found skipping out of the limelight and acting as a kind of chattering woodwind decoration.

The second subject of this movement is in ideal contrast to these rather precisely shaped fragments. It is a long graceful singing tune which first appears on the 'cellos and whose Russian character is unmistakable:

211

After a restatement of Ex. 210, the exposition ends quietly and then, back in 3/2 again, the highly-charged and finely controlled development begins with the timpani setting a simple but irresistible rhythm which is fundamental to the rest of the movement except when the music moves back into duple time. The recapitulation is again no simple restatement and contains further expansion and fresh ideas.

As in all Borodin's symphonies, the scherzo is placed second and, like that in the First symphony, is a one-in-the-bar affair, bearing this time the unusual time signature of 1/1. Nor does the similarity between the two movements end here. Again it is marked *prestissimo* and again there is a trio middle section, although not this time actually called such. But more remarkable still is the affinity between the melodic material of the two scherzos. In the Second symphony the principal elements are again rising and falling *staccato* figures and a four-note sequence (ascending this time) which will be instantly identifiable and of which much use is again made. There is also a more sustained off-the-beat figure. But again it is in the middle section that we are more aware of Borodin's individuality. Here the overtones of *Prince Igor* are unmistakable as the oboe introduces this characteristic lilting melody:

212

The key of the *Andante* is D flat major. After a three bar introduction on the clarinet over harp chords, the horn sings this long, rather lonely and highly evocative main melody:

213

In addition there are a couple of other short phrases which Borodin also introduces to fruitful effect. One, a dark rising and falling figure, is also first heard clearly on the horn though in fact it comes at the end of the clarinet version of the main melody. The other, more prominent, is the slightly wistful one-bar phrase which the oboe then introduces. This material Borodin moulds into a movement of great and often intense beauty, moving with consummate ease from rather withdrawn reflection to almost theatrical declamation until the main melody is presented in a rich *tutti* of brilliant orchestral colour. Then the movement dies away to end as it had begun with soft clarinet and horn solos over the lightest of accompaniments. It is a beautifully shaped movement, highly concentrated and yet relaxed, and sustaining throughout a moving personal and national conviction.

It leads without a break into the finale, a vigorous B major *Allegro*. The relentless rhythm of the opening clearly belongs to the same world as the Polovtsian dances in *Prince Igor*, to which the first subject is also closely allied in spirit. It is heralded in a magnificent *tutti* of alternating 3/4 and 2/4 bars by the violins and upper woodwind, against whom certain of the lower instruments have a broader rising figure while most of the brass complete the harmony. It has thus both a glittering panache and a rugged strength which together are irresistible. The second subject when it follows is a complete contrast, although it too evokes the atmosphere of *Prince Igor*. There is surely a close affinity between it and the languorous song of the maidens in Khan Kontchak's camp:

214 

Despite its clear origins in the dance and its apparent sympathies with the spirit of the theatre, this movement is orthodox in shape although the detailed procedures are entirely personal. The orchestration fuses brilliance and power in a magnificent way, and the whole effect is one of great strength of purpose allied to complete mastery of means so that this movement emerges as a worthy finale to what is both a highly evocative and exciting work and one of the great symphonies of Russian literature.

Borodin's unfinished Third symphony is rarely played, but it has recently begun to receive some attention in Britain, and so a brief comment may be helpful. It is in A minor and its two movements were only sketched by Borodin; they were completed and orchestrated by Glazunov.

The mood of the first is in keeping with its restrained economical orchestration, for it is far more delicate than either of its counterparts in the other symphonies. The first theme, on unaccompanied oboe, has, for example, a dancing folk-song character, very beguiling but hardly material for a heroic structure. It does, however, effectively set the course of the agreeable, good-humoured and unproblematic movement which follows.

The D major scherzo is also essentially straightforward and light in texture (the trombones, for example, are excused attendance), but though uncomplicated, it is nonetheless unerringly well-calculated. Again there is a contrasting central trio, which consists of music originally intended for *Prince Igor*, easily recognizable as such. The flanking *vivo* sections are enchanting and by and large more personal documents than either of their counterparts in the other symphonies. In 5/8, with occasional neat changes to 2/4, the whole thing has a delightful lilt and simple gaiety as the main melody, first heard on the oboe, should indicate:

215

13
JOHANNES BRAHMS
(1833–97)

Julius Harrison

*

In this age of turmoil and iconoclasm Brahms's symphonies still stand four-square to the world: pillars of classical architecture on whose firm, consonant foundation nineteenth-century romantic sounds soar upward in a preconceived plan mindful of every detail however small. Nothing is left to chance; each movement has its course determined from the very first note. Imaginative and original themes, individual harmonies, rhythms and phrases set in unusual metrical patterns, complete mastery over counterpoint, orchestration consistently serving the spirit of the music without flamboyant rhetoric, all are to be found within a classical framework controlling everything. Further, the wide range of each symphony can be said to stretch from the stolid burgher-like milieu of the composer's native Hamburg to the warmly vibrant, mellow sounds of his adopted, Schubertian Vienna where he lived for so many years. Then to all these facets of his art must be added an ever-recurring sense of melancholy, of brooding loneliness; hints of autobiography shyly concealed; and, by way of contrast, a love of joyous gypsy sounds which, through his early association with the itinerant fiddler Reményi, made him arrange or compose Hungarian dances.

Brahms was certainly one of music's paradoxes, from early years carrying an old philosopher's head on young athletic shoulders; playing something of a lone hand in an age of violent change – the age of Berlioz, Liszt and Wagner. Such was his integrity, his loyalty to the past, that at times in his music he came near to being a pedagogic bore. Never in his large-scale works would he permit himself any easement from those principles of form, counterpoint and development so firmly founded on Bach and Beethoven. Form was to him the *fons et origo* of all he wrote, even to the extent of driving or coaxing to their sternly logical conclusion insignificant thematic phrases amounting to little more than the small change of music. This, for example, we can hear in the 250 bars which comprise the *Adagio non troppo* movement of the D major Serenade, composed in 1858, when Brahms at twenty-five knitted his music together like some old body diligently plying her needles.

Yet within this thoroughly German concept of his art, within the somewhat restricted limits with which he saddled himself, he succeeded mightily in his efforts to unite classical forms and nineteenth-century romantic expression. Even such a loose term as Rhapsody meant to him a disciplined form controlling subjective emotional utterances and keeping him poised continually somewhere between restraint and the ever-present impulse to create music revealing his inmost feelings. Maturity came to him at an early age. As Richard Specht wrote in his life of Brahms, 'he was at the age of twenty the same finished, self-reliant human being as at sixty, and as a creative musician too he could only grow more perfect, but not more highly endowed than he was in his youth'.

Of that Brahms was himself aware. Looking back over the years, he once ventured the opinion that in his music he seemed always to be 'milking the same udder'. This to a great extent is true; he never rushed headlong into passionate displays of anguish or high spirits. He and his contemporary Tchaikovsky were as the poles asunder, each making his own significant contribution to a symphonic form now, in their lifetimes, charged with all the flexibility demanded of it since romantic and iconoclastic practices had crept in to take a hand in shaping things more fantastically, and, on occasions, almost unrecognizably. The symphony had for years parted company with many traditions established before the time of Mozart and Haydn, leaving Brahms a solitary figure on a hard-cobbled road of his own choosing, true heir to Beethoven as he tried to be; constantly reminded, he once told the conductor Hermann Levi, 'of that giant whose steps I always hear behind me'.

To say that he improved on Beethoven or even challenged him would of course not be true. Rather, he saw and heard increasing possibilities of symphonic developments in a manner not fully explored by his precursor: new key-relationships in and between movements; new and intriguing chromatic harmonies incorporated solely to enhance the effect and prove the strength of the fundamental diatonics that remain the crowning glory of Beethoven's and his own music. 'Beethoven's No. 10' is not altogether an inapt description of Brahms's First symphony, however much the actual musical speech of the two composers differs in accent, dialect, idiomatic expression or what you will. As we shall see later, Brahms in three of his four symphonies, created meditative third movements the very antithesis of Beethoven's ebullient scherzi, using a smaller orchestra to paint water-colours or pastel shades in complete contrast to the large canvases of his outer movements.

How he managed to harness his technique to undeniable inspiration, rarely failing to control steed and chariot, is one of those mysteries which (as with Bach, Mozart and Beethoven) defies explanation. He is certainly in the company of the few, whether in his symphonies or in his other works. Yet in spite of these manifestations of his peculiar genius it should be remembered that the four symphonies have not met with entire acclaim. The marked restraint (for the times) in the matter of their orchestration has proved for some critics something of a hindrance to full appreciation. Comparisons with the richer styles of Bruckner and Mahler, with the earlier, fantastic Berlioz, take us nowhere. Brahms must be judged solely on his own merits, even when he so methodically counts over his small change, fascinated like some numismatist by the obverse and reverse sides of his coins.

Settling in Vienna in 1863, where through his association with many famous musicians, writers, poets, painters and philosophers he soon became much more than an imported figurehead in the city's artistic life, he still brought with him, in symphonic music to be composed there or elsewhere in Austria, the pungent tang of the North, and a somewhat evangelical outlook not to be repressed however much he might in lighter moments surrender to the spellbinding joys Vienna's volatile life had to offer. Brahms and Bruckner, the devout Catholic, were thus, at the opposite points of music's compass; both living in Vienna at the same time; both dying within months of one another; both geniuses of widely differing orders whose masterpieces had scarcely anything in common.

The instruments Brahms used in his symphonies were chosen with infinite care. Except for the piccolo which replaces the second flute in the third movement of the E minor symphony only the customary woodwind are employed. To these, in three out of the four symphonies, a contra-bassoon is added, often with telling effect. In the Second symphony the bass tuba takes its place. Trombones are used sparingly: in the First and Fourth symphonies only in the finale; in the Second and Third in three movements out of the four. The triangle, entertaining us merrily in company with the piccolo in the third movement of the Fourth symphony, is Brahms's only use of extra percussion. In his use of horns (two pairs) and trumpets he stuck mainly to that limited series of notes which the old valveless instruments were able to play. Thus he met the classical usage of these instruments more than halfway, and chiefly through them preserved a fundamental diatonicism.

Another feature was the wide choice of keys for the various movements: C minor, E, A flat, C minor and major for the first symphony; D, B, G, D for the second; F, C, C minor, F minor and major for the Third; E minor, E major, C, E minor for the Fourth.

It will be seen that although Brahms will always remain a somewhat controversial figure in certain aspects of his art, he, like Beethoven, was an eminently practical symphonist who was a master of design, able within that design to coordinate ideas, their varying tonalities and their orchestral expression in a way few composers have equalled.

No definite programmes, no illustrations of personal experiences in life itself, no recourse to pictorial imagery outside the bounds of absolute music can be found in the four symphonies. Only that very personal motto-phrase on which the Third symphony depends so much – the three ascending notes F, A flat, F ('*Frei aber froh*': 'Free but glad'), which Brahms adopted when a young man and afterwards used frequently in his melodic outlines* – rings loudly through the music like some cry of the heart wrung from him as he contemplates the passing of the years.

Brahms spent at least twenty years on his First symphony before it was given its first performance in 1876 – the memorable year which also witnessed the production of Wagner's *Ring of the Nibelungs* at Bayreuth. To analyse it fully; to expatiate on the dexterous use of counterpoint; to describe in lavish detail its moving harmonies or its highly-organized orchestration, all are beyond the scope of this present article. The reader who desires more explicit information is therefore referred to the first volume of Tovey's *Essays in Musical Analysis* (O.U.P.) in which all four symphonies are analysed at length in penetrating programme notes written over sixty years ago for London performances by the Meiningen Orchestra†.

This First symphony is, like Beethoven's Fifth, a symphony of conflict and triumph: the resolving of an intense C minor struggle into a concordance of C major sounds uplifting and victorious. The first movement, Max Kalbeck stated in his biography of the composer, was the outcome of Brahms's deeply-felt emotional reactions to Schumann's tragic end in 1856. If that be so, we can well understand how for many

* Not always at the same pitch. Cf. the second of the four Ballades for piano Op. 10; the first movement of the D minor piano concerto Op. 15; the song *Minnelied* Op. 71, etc.

† And to Julius Harrison's own book *Brahms and his Four Symphonies* (Chapman and Hall, 1939). (Ed.)

years Brahms found himself reluctant to complete a symphony as personal to himself as it was to Clara Schumann who meant so much to him. Nonetheless, the music transcends every local or topical circumstance. It is absolute music in its own right; one of those symphonies that, in Hadow's words (in another context) 'stands to Music as the epic to Poetry'.

In the *sostenuto* 6/8 introduction Brahms not only foreshadows the main themes of the passionately striving *Allegro* which follows, but through them already establishes a mood of extreme tension that persists throughout much of the symphony, even though the *Andante* movement and the *Allegretto* movement (in lesser degree) bring with them some temporary respite from the struggle. In Ex. 216 violins and 'cellos lift up their heads confidently, while pitted against them woodwind, violas and horns do all they can to drag the music down in collaboration with timpani, contra-bassoon and basses pounding away at reiterated Cs at their lowest pitch.

216

It is this alternation between the spirit of combat and that of despair which makes this *Allegro* movement such a fascinating study in musical psychology. While it is impossible to quote even some of the most important themes in all their combinations, Ex. 217 indicates how charged the music is with these opposing elements.

217

Thus through Brahms's inexhaustible contrapuntal skill the vigorous phrase (*a*) becomes inverted at (*b*), here indicating with something like savage intensity, particularly near the close of the movement, that the spirit of defeat looms over everything like the blackest of clouds. The same inverted treatment of the first three notes of Ex. 216 (C, C sharp, D on violins, etc.) is also a marked feature of the later development section. At this point the three notes drop in pitch and volume till they sink to the very nadir of hopeless expression, from which ascending once again in new-found strength to a remarkable climax, they tell us that the fight is by no means lost. Even so, in a fierce onslaught, the first two notes of Ex. 217(*b*) finally hammer the music mercilessly to defeat. Ex. 216 in a *meno allegro* tempo corresponding to the *Sostenuto* introduction, ends the movement in a pale reflection of the epening bars and of Ex. 217(*a*). In these retrospective moments partly reiterated Cs on timpani and horns add their now hesitant contribution. Though the music certainly ends in the major key, it suggests no more than a temporary lull in the conflict.

The lyrical *Andante sostenuto*, pitched a major third higher in E major, stands in the same relationship to the first movement's C minor tonality as does the second movement of Beethoven's third piano concerto. Thus we see Brahms exploiting, not only here but throughout his four symphonies, chromatic key-relationships of a more romantic order; something that brings to each symphony new colours, new shades of feeling, freshness, vitality and contrasts no sensitive ear can fail to appreciate, even were the composer's deft planning not fully grasped.

The main theme of this movement is indeed beautiful.

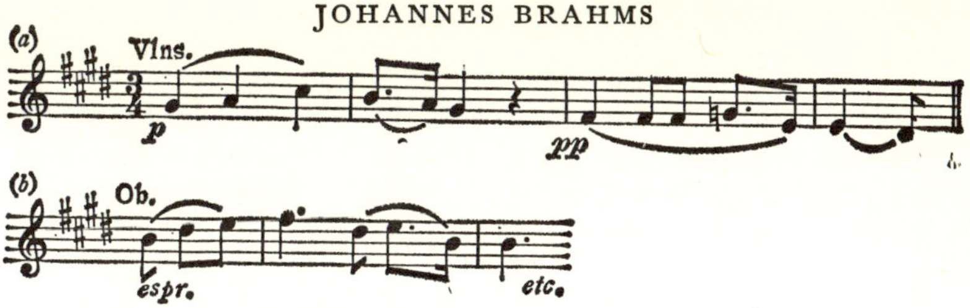

218

Its first strain (*a*) mostly on strings, spreads over seventeen bars; its second (*b*) contains an equally appealing oboe solo against whose last two bars Brahms makes use of the opening notes of the main theme by way of string accompaniment. From this point the music develops in agitated, syncopated style with another solo passage on the oboe. It is then followed by one for the less poignant clarinet, sounding like a first ray of light in music which has by now reverted to the more disconsolate mood of the first movement. When the recapitulation of Ex. 218 is reached, the orchestration is entirely different: full and more intense in general expression. Then in a prolonged coda that recaptures the serene spirit of the opening music, the oboe – Ex. 218(*b*) – is joined by a horn and a solo violin in extended romantic passages which eventually lead to the peaceful close. Brahms confessed he had made a cut in this section. As it now stands, it seems in right proportions to the movement as a whole, charged though it is with a romantic feeling which, like Juliet's, makes parting such sweet sorrow.

Turning to the key of A flat for his third movement Brahms once again raises the pitch a major third, G sharp becoming A flat. (For the finale he completes this ingenious scheme of key-relationships by returning home to C minor and then C major.) Though the movement corresponds in form to a scherzo and trio, it is in spirit more like an interlude, placid and semi-agitated by turns, not far removed from chamber-music in its general character, a study in melodies uncommon in their various shapes, all beautifully harmonized and avoiding those elaborate contrapuntal developments which make the finale such an overwhelming display of Brahms's inspired intellectual powers. Truly we can say that Brahms was 'saving up' for this finale by restraining the third movement. Yet he is ingenious. The clarinet's opening theme has two five-bar phrases, the second of which is an exact inversion of the first.

219

To this the woodwind respond with a new theme of eight bars, its metrical quantities divided into phrases of 2, 2 and 4 bars. Consequently the rhythmical outlines of this movement continually shift from one shape to another. But that is not all. When the violins, at the conclusion of the woodwind response, repeat Ex. 219, they extend the long notes by two more bars – shown above as (6, 7). Gently disturbed developments now occur, aided by fragmentary new phrases making their solitary appearance in the movement. The middle B major section is in vigorous 6/8 time and comprises lively exchanges of ideas between woodwind, brass and strings. (Timpani have no part in this movement.) These exchanges, in formal trio style, lead to the return of Ex. 219, somewhat altered and now divided between clarinet (five bars) and violins (six bars), the music being given yet another shape. But Brahms, wasting no time over formal repetitions, soon reaches his coda. In their subdued mood these final bars seem designed to prepare us for what is to follow. Phrases once so vigorous in the B major section now become inert and dark in colouring in what is nevertheless a beautiful coda: one admittedly melancholic in Brahms's own way, yet one that demonstrates convincingly that unfailing unity of thought which, like some bridge, carries this movement over to the next one.

The finale has an extended, intensely dramatic C minor

and major *Adagio* introduction foreshadowing almost every theme or phrase with which the subsequent *Allegro* movement is concerned. Nothing like it had been written since Beethoven's Ninth symphony. In struggles towards the light out of some sort of primeval darkness, full of 'human terror and expectation' (Tovey), yet so confidently that we can sense the outcome long before it happens.

To quote every theme or phrase in its embryonic or complete form is not possible here. But in Ex. 220 it will be seen at (*a*) how the main theme of the *Allegro* (*b*) is approached. Only at (*b*) is the keynote of the now triumphant music fully established with a glorious C major theme sixteen bars long.

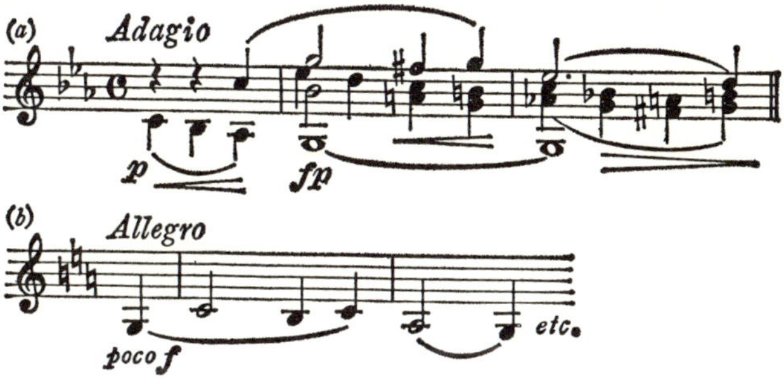

220

Where, still in the introduction, the music turns into C major at the *pp* entry of the trombones, a solo horn bursts in with a great theme, Ex. 221 (*a*), one derived from an Alphorn melody Brahms once heard.

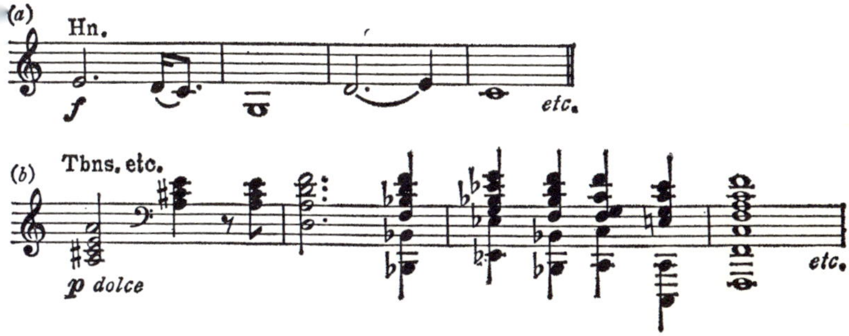

221

To this a solemn chant-like response, Ex. 221 (*b*), now comes from trombones, bassoons and contra-bassoon to complete the material which the *Allegro* movement is to develop in symphonic fullness and continuity from the point where Ex. 220(*b*) is first heard. From now on the music gathers strength in remarkable displays of contrapuntal inspiration. Its sheer optimism, its unerring sense of direction, of a goal to be reached through final triumph, all are co-ordinated. In the ecstatic *più allegro* coda, during which the solemn chant of the trombones, Ex. 221(*b*), now returns in a glorious *fortissimo*, the first violins eventually soar to their highest possible C. In that fact, small though it may seem compared to other things, we are given final proof that this great symphony was indeed intended by its composer to aspire to heights beyond the reach of all the gloomy and even sinister C minor sounds with which it began.

Composed within a year after the completion of the First symphony, No. 2 is in its general style and outpouring of happy ideas the very antithesis of the other. Brahms spent the summer of 1877 at Pörtschach in Carinthia; described by him as 'replete with Austrian cosiness and kindheartedness'. There he worked on his new symphony and into it infused a sunny warmth, a genial spirit, which, after the First, took his critics completely by surprise. It all 'sounded so merry and tender, as though it were especially written for a newly wedded couple' – such was his own description.

The first movement begins in the simplest way. Three notes, D, C sharp, D, seem at first hearing no more than a preludial figure designed to support a homely theme on horns and woodwind starting a bar later on F sharp – Ex. 222(*b*).

222

But these three notes mean much more. Both as a theme in itself and as an almost ubiquitous accompaniment (in crotchets, quavers and even semiquavers) they govern the whole movement (and indeed the finale), varied though their pitch will become as the movement proceeds. In the development section the three trombones declaim them strenuously in close imitation; in the lovely serene coda near the end they are given inspired treatment, packed playfully together in conjunction with brief references to the other main themes, Ex. 222(*b*) and Ex. 223. And still another theme of equal importance is included in this calm opening. Starting on the note A on the first violins, Ex. 223, lyrical in mood with its flowing quavers, completes a trinity of interrelated themes.

223

Thus from the *initial note* of each theme – D, F sharp, A – Brahms builds up an implied triad-chord of D major which ultimately, on *ff* trombones, reaches its predestined end in the final bars of the symphony. Other themes, notably a warm-hearted melody started by the 'cellos, together with strenuous cross-rhythmed figurations, complete a first movement as romantic in its speech as it is classical in design. Towards the end the first horn has a strikingly beautiful passage – the longest Brahms ever wrote for that instrument. Here indeed is the magic of a summer evening; the music of a tranquil mind at one with nature, no longer concerned with psychological problems crying out for solution.

A characteristically solemn mood follows with the B major second movement; the 'merry and tender' sounds of the first movement certainly demand a well-defined contrast. Even so, the first note of the broad 'cello melody, F sharp, serves as a total link between the two movements – just as this movement's final B will determine the G major key of the third.

224 

Against the twelve bars of this highly original 'cello theme plaintive sounds issuing from bassoons, and particularly from trombones in harmony, create a darkened mood which is to prevail throughout the entire movement. Three other themes follow to complete the main subject-matter of this section: one for horn, closely imitated by oboes, flutes, 'cellos and basses; a second syncopated style for flutes, oboes, clarinets and then violins; and a third, which, after a momentary silence, starts simply enough on the strings. But as Tovey puts it: 'A child may say a word which makes history; and so this unpretending theme startles us by moving, with a rapid *crescendo*, into distant keys, and blazing out in a stormy *fugato* with a counter-subject in flowing semiquavers.' This is the middle section; it travels through various minor keys in a disturbed state which affects even the first few notes of the 'cello melody, Ex. 224, when now given to violins and then oboe. In the recapitulation that follows, this theme and what grew out of it originally are orchestrated quite differently and with no little emotion when they reach a new point of climax. Here agitated semiquavers on violins and violas play restlessly round those woodwind and brass harmonies (now in the minor) that were such a feature of the opening bars. The movement ends with broken phrases from Ex. 224 – inflexions of B minor harmony darkening the music before the close in the major.

In this otherwise carefree symphony no more complete change of style could be imagined than that which makes the *Allegretto* movement such an endearing sequel to the disturbed *Adagio*. Hardly a scherzo proper, more like an intermezzo, it is at once playful and ingenious. An oboe solo, later shared by clarinets and flutes and set against a background of *pizzicato* 'cellos, bassoons and a horn, serves as a dulcet theme, which in two subsequent and widely differing *Presto* sections is so metamorphosed – Ex. 225(*b*) and (*c*) – and yet so close to the original no matter how disguised, that we can only marvel at Brahms's unfailing resourcefulness. Like Prospero's directive to Ariel, he could have said here: 'Thy shape invisible retain thou still'.

225

The form of this movement is somewhat akin to an unorthodox rondo, for the 3/4 *Allegretto* returns after each *Presto* section, though altered in many details. In its changeable rhythms, its exquisite harmonies and modulations, its gracious *Allegretto* moods and melodies, its contrapuntal feats, its strange dalliance between major and minor keys (those of G particularly), its avoidance of the heavier instruments (trumpets and drums), this movement is a perfect congruence of delicate sounds.

When W. S. Landor once wrote that 'the present, like a note in music, is nothing but as it appertains to what is past and what is to come' he certainly touched on a palpable truism that might well apply in a still wider sense to every symphony worth the name. Here in the finale of his Second symphony, Brahms takes a quick look back before proceeding on his way, returning, now in broadly flowing duple tempo, to the happy mood of the triple-time first movement, while imparting to his new music such rhythmical vitality and cumulative strength that the whole work moves inevitably forward, crowned in the end with D major splendour.

The movement begins *sotto voce* on strings alone (Ex. 225), its opening bar recalling the D, C sharp, D of the first movement, and (rather more obliquely) the F sharp of Ex. 222(*b*) and the A of Ex. 223, separated though these notes are by others intervening.

226

Yet it should not be forgotten that this finale has its own individual thematic life quite apart from that of the first movement. After the twenty-two *sotto voce* bars with which the movement starts, and which contain, incidentally, close on the heels of Ex. 226, an equally important theme mostly composed of descending fourths, the music bursts without warning into a state of festive incandescence rarely to be extinguished throughout the movement. But when this major-key brilliance does abate from time to time, more especially in the development section, Brahms turns to darker passages in the minor keys that not only create the right symphonic contrasts, but also give him opportunities for indulging in the most dexterous contrapuntal feats. Themes are inverted, combined, or shaped anew. The first bar of Ex. 226 has its quavers smoothed out into six crotchets to the bar; the important thematic phrases with their descending fourths, originally crotchets, become sleepy minims when trombones and tuba steal furtively into the movement for the first time. All these feats of inspired craftsmanship, never artificially contrived, lead to the recapitulation and to the intensely dramatic coda. There the broad second main theme of the movement (first heard in A major) now returns in the key of the symphony, compelling and triumphant.

227

But its career is by no means over, for in the magnificent peroration, Brahms reintroduces its first notes and harmonies, now in the *minor* key, on trombones, tuba and woodwind, modulating from D minor rapidly through C major to B flat major. Here the whole orchestra gives Ex. 227 ecstatic utterance. From this point to the end of the symphony Brahms gathers all his main themes together in a truly remarkable concatenation of joyous sounds. And when the trombones, five bars from the end, proclaim *ff* their D major triad, we know that what was promised in the early bars of the first movement (Ex. 222 and 223) has now come to its predestined fulfilment.

Returning to Pörtschach in the summer of 1878 Brahms wrote to Eduard Hanslick the famous critic: 'So many melodies fly about, one must be careful not to tread on them.' Such was his whimsical postscript to the many 'merry and tender' melodies which make this Second symphony proof against every change of fashion to which the art of music is so perversely susceptible.

Brahms was fifty when he completed his Third symphony in 1883. Romantic as ever in his themes and harmonies and still faithful to classical forms, he yet brought to the music something quite different in its spirit and general style. Gone are the stricken moods and final triumph of the First symphony, the homely beauty and exhilaration of the Second. Instead, there is now more than a suggestion of a composer rebelling against the march of the years; a still youthful heart trying to come to terms with middle age creeping on all too soon.

As a result of these contrasting moods Brahms created

another masterpiece that in its essence spans thirty years of his life. Taking the youthful motto-phrase of 1853 with its three ascending notes – F, A (flat), F – to which reference has already been made, he made this not only the servant of a still more important theme, but also on occasions gave it pride of place, harmonizing it so diversely in the first movement that between it and the main *passionato* theme the closest integration was achieved. In Ex. 228, for example, it will be seen how closely these three notes cling to the main theme.

228

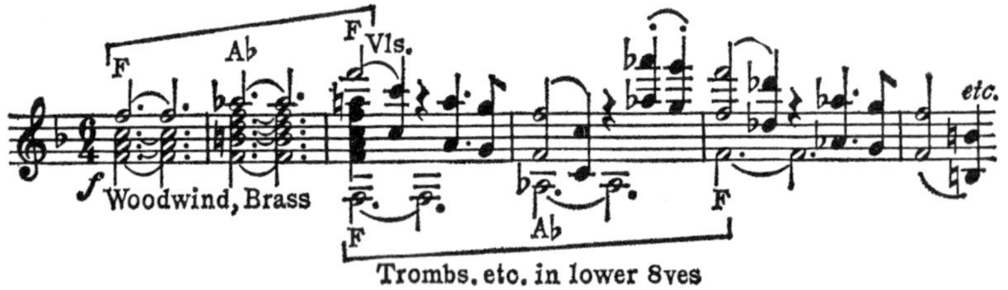

And as the *passionato* theme modulates to various keys, so does the motto-phrase change the pitch of its notes. Then in the key of A the clarinet plays a new and lovely theme over a drone bass on violas and 'cellos – a pastoral theme in 9/4 time which gets in a veritable tangle of notes, and is soon followed by playful woodwind exchanges in subtly-contrived cross-rhythms. The motto-phrase returns on the oboe, and from this point the music is subject to a continual flow of quavers leading in growing excitement to the repetition of the first part of the movement. Two things stand out in the development section: agitated minor-key versions of the pastoral clarinet theme, now given to strings and bassoons; and then a truly magical entry of a solo horn playing the motto-phrase, first in E flat and then in G flat. Shadowy references to the main *passionato* theme – note the contra-bassoon – now foretell the imminence of the recapitulation. Much of the earlier music returns, but in other keys. Finally, Brahms lets himself go in a magnificent, gripping, sinewy coda section where the main theme, together with its companion motto-phrase, is given more passionate utterance than at any other time. In a riot of contrapuntal devices all glowing with an inner fire we are reminded of that struggle between youth and age in which Brahms seems so involved in this symphony.

The C major *Andante* is serenely beautiful, interrupted but once by a fortissimo outburst lasting no more than half a bar. Clarinets and bassoons carry the main burden of the first section with thematic phrases varied at each entry, while interspersed echoes on violas and 'cellos vaguely recall the motto-phrase, even if its third note does fall short of the octave.

229

As the movement develops, strings add semiquaver movement to Ex. 229, but not for long. Then a dark cloud steals over everything by way of the four bracketed notes marked (*a*) in Ex. 229. Out of it emerges a new, subdued theme, one that not only plays a very important part here, but also is destined profoundly to affect the whole structure of the finale.

230

With its treatment of the motto-phrase and the main theme of the first movement (Ex. 228), both of which appear in the finale, the Third symphony, then, belongs to the 'cyclic' class, put together in no haphazard way, but with all its elements and motifs welded into an organic whole. To describe the development of the second movement in more detail is not possible here. But in a supremely moving coda section on which twilight descends, the ever-changing chromatic harmonies founded on the first two notes of

Ex. 230 bring with them a sense of peace not broken till the finale is reached.

But even if there is nothing to ruffle the mood of the third movement, the music now turns, pensive and even melancholy, to the key of C minor. Once again, while adopting the ternary form of scherzo and trio, this movement is no more than a brief interlude with a haunting melody shared successively by 'cellos, violins, horn in company with flute and oboe, and again (after the middle section) by horn, then oboe, and finally by violins and 'cellos.

231 

Its six appearances suggest that Brahms regarded this melody with lingering affection, decorating it with fanciful arpeggio figurations, mostly on strings and sounding like fine tracery. No trumpets or drums disturb the mood. The middle section of the movement, wrought so skilfully, shows that Brahms took infinite pains to transmute a three-note phrase of no great consequence into the gold of applied craftsmanship. Only two brief but tender phrases for strings break the three note patterning here. But taking the movement as a whole, its grave beauty is undeniable. Static in mood, it is a poignant interlude before the storm of the finale breaks upon us.

Here is a movement as inspired as anything he ever wrote. Set in the key of F minor and only melting into the major key in the concluding forty-three bars, it is a remarkable tussle between opposing musical elements. The opening theme, mysteriously furtive in its quavers, contains in one form or another phrases that will be given the most malleable treatment in company with Ex. 230 of the second movement. Ex. 232(*a*) shows the opening bars, and at (*b*), above them, an ingenious 'skeleton' version that will soon play a most important part in the development of the music.

232

But shortly before this happens, Ex. 230 steals in *pianissimo* on trombones, strings and wind, yet so far only in a tentative way. Its great moments are still to come: first in the passionately emotional upheaval in the middle section, and then in the serenely beautiful F major coda with which the symphony ends. When Ex. 232(*b*) suddenly bursts in on full orchestra the mysterious mood of the opening bars is swept aside by many stormy passages which from now onwards will dominate everything while the music remains in F minor. A swinging, cheerful 'second-subject' theme follows on 'cellos and horns. Then the motto-phrase reappears, but is soon lost in the general tumult. In due course tension is temporarily reduced as Ex. 232(*a*) returns in various forms. And from its ghost-like phrases Brahms now builds up his magnificent central climax founded mostly on Ex. 230 of the second movement. Development and recapitulation merge into each other with perfect dovetailing of all the main themes. Then as the stormy elements subside, Brahms softens the edges of Ex. 232(*a*), turning its quavers into six-to-a-bar crotchets on muted violas etc., and then into still longer notes when the oboe takes the theme up at that point where the F major coda of forty-three bars arrives. The strings (violins now muted), re-introduce the semiquaver arpeggios heard near the opening of the finale. Against their continual flow brass and woodwind take up the now familiar Ex. 230 of the second movement, giving to its exquisite harmonies the colours of a wonderful sunset after a day of storm. The motto-phrase returns hesitatingly in the final bars, followed by *pp* descending tremolo semiquavers on the strings which echo in the far distance the *passionato* theme of the first movement. Everything is thus resolved in this tranquil close.

For the most part austere in character and less given over

to romantic expression, the Fourth symphony marks the culminating point in orchestral music of Brahms's genius. It has a classical grandeur. None of its attributes, its themes or harmonies lends itself to pictorial illustration; we can interpret each only in terms of music absolute in its own right. Generally accepted as his finest symphony, the Fourth stands apart, aloof in a world of abstract thoughts, relieved only by the *Allegro giocoso* third movement Brahms wrote after the three other movements were completed.

The symphony belongs to 1884–5. Like his friends and critics, Brahms was full of misgivings about it: himself not on account of the music, but more in dread of its reception by the public. Hanslick, staunchest of Brahms's supporters, described his reactions to the first movement as 'the sensation of being cudgelled by two terribly witty men'. In the light of close analysis, as the various phrases of the long opening theme develop, such early reactions can well be understood. For near the end of the movement this theme precipitates a battle-royal between *augmented* and *diminished* intervals (mostly fourths) which, on musical grounds alone, is without parallel in symphonic art. The extreme tension thus created must at first have been baffling.

This long opening theme (Ex. 233) must be quoted almost in full, for within it there are three definite subdivisions (marked *a*, *b*, *c*) each of which is subjected to exhaustive treatment. Nothing goes to waste, not even the <> in (*b*) or the quaver rests in (*c*).

233

Within this tripartite scheme Brahms actually creates three distinct moods. That of (*a*) is here quietly ruminative, yet by no means so as the movement unfolds, for it is often ingeniously varied (woodwind quavers etc.) and then, in the end, is thundered out in closely-knit imitations by the full orchestra – an upheaval which creates an even greater one through the intense struggle between augmented and diminished intervals. The second section (*b*), while given similar treatment and becoming as active as (*a*), is most remembered in a series of hushed phrases occurring just before the recapitulation. Here a succession of soft chromatic harmonies take the music into another world – a spirit world of eerie < > sounds. Section (*c*) with its opposing intervals, its agitated phrases broken by quaver rests, supplies strongly contrasting rhythms which near the end of the movement assume a degree of violence that sweeps everything before it. Equally important are other subsidiary themes with the same characteristic augmented or diminished intervals.

234

Even when other notes intervene, as in Ex. 234(*a*), the resultant effect is still one of extreme tension. Ex. 234(*c*) is a brief extract from the bold theme which, in academic terms, can be loosely described as the 'second main subject' – a sonata-form term which in all symphonic movements of many themes has but ambiguous meaning. Even this theme is affected by the prevailing discordant intervals. Needless to say, there are moments during the course of the movement when the music relaxes in the major key; but never for long, for Brahms seems possessed throughout with his discordant elements – how to pit them implacably one against the other, how to contain them within an orthodox form, and how finally to resolve them within the concordance of his E minor key. This last he does with an austere plagal cadence (A minor to E).

No less unusual is the E major second movement. While it cannot be said that it lacks romantic feeling, especially in the lovely 'cello melody which grows thematically out of strongly marked semiquaver triplets (see Ex. 236 below), yet there is a whole world of difference between it and the E major slow movement of the First symphony. Here a more profound mood: remarkable juxtapositions of the Phrygian mode* and the key of E major bring to the music a strange blend of old and new – *Et nova et vetera* might well be its motto.

The Phrygian opening is given to horns with woodwind added at the second bar.

235 

Halfway through the fourth bar the clarinets steal in *pianissimo*, joined at the fifth bar by *pizzicato* strings, the key of E major now established, though with G, D and C naturals (the Phrygian version) as a prominent feature of the harmony. These early bars of the movement create a strangely remote atmosphere. But soon the music leans more and more towards E major proper as it develops its warmly romantic phrases before modulating to B major for the deeply felt 'cello melody. This melody is forestalled in vigorous fashion by triplet semiquavers shared between woodwind, horns and strings.

236 

* The scale of eight notes starting on E corresponding to the white notes on the piano.

Out of these contrasted elements the music is developed and then recapitulated with marked changes and different orchestration. Ex. 236(*a*) is given more passionate utterance just before the 'cello melody returns on violins, tenderly and then strangely syncopated. In the coda, in familiar Brahms style, mysterious shadows steal across the music. Brahms is at his finest here; no other composer ever surpassed him in the art of *chiaroscuro*. The last six bars of the movement are remarkable in that both the Phrygian and E major versions of the opening theme become interlocked in a convulsive effort before reaching a most unorthodox final cadence – F major to E major, all over a pedal E.

The C major third movement – Brahms's nearest approach to a truly symphonic scherzo – is not only playful, but masterful as well. In the context of this otherwise intensely profound symphony it sounds startlingly extrovert. Once again, its main theme contains three subdivisions, here illustrated in Ex. 237(*a*), (*b*) and (*c*).

237

All are given the most vigorous rhythmical treatment, controlling the whole movement even when their intense activity leads to other episodical themes of short duration. Brahms's play with invertible counterpoints is everywhere in evidence. Thus Ex.237(*a*) is often turned upside-down, the lower part taking a prominent place in the thematic development, especially when the movement reaches the remote key of C sharp minor. Here the music begins to lose its forceful character as it moves towards a tranquil middle section in D flat major, where, as Elizabeth von Herzogenberg, writing to Brahms, said of it aptly: 'all the gay apprentices slouch home from work and the peace of the evening sets in'. But not for long. Ex. 237(*c*) bursts in suddenly in its original key of E flat. Much of the earlier part of the movement is now repeated, and with increased vigour through the semiquavers shown in Ex. 237(*b*). Finally, in an exhilarating coda, trumpets and horns turn Ex. 237(*c*) into a C major fanfare, affirming the gaiety of this irresistible movement.

The mighty finale is founded on a *chaconne* bass which Bach used for the last chorus in his 150th church cantata. Brahms took this skeleton bass, altered the note-values and introduced a chromatic A sharp to create an astringency without which, as he told the conductor Hans von Bülow, everything would sound 'too lumpish, too straightforward' for a purely orchestral work. It is therefore interesting to compare Bach's original with Brahms's expansion of it into a theme of eight bars, each note now being of equal length and the key changed to E minor.

238

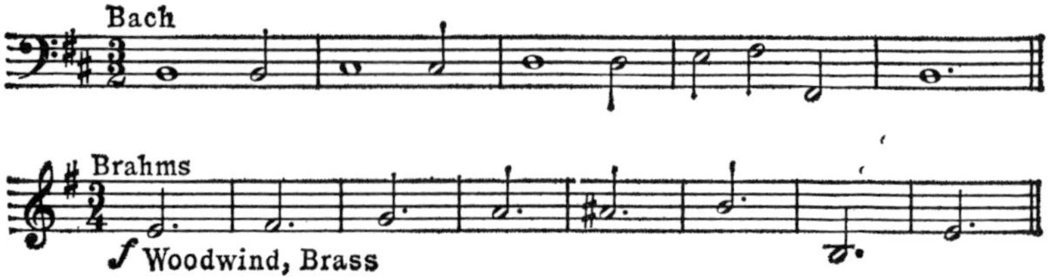

On this theme, its eight-bar periods never broken, Brahms constructed a series of thirty variations* and a magnificent coda which not only rounds off the movement, but also serves to complete the grand architectural design of the whole symphony. The variations explore intricacies of counterpoint, tortuous labyrinths of chromatic harmony and the whole range of orchestral expression in such a way that the eight-bar divisions seem scarcely noticeable as such.

The theme is announced starkly on brass and woodwind, the trombones making their first appearance in the symphony. Twelve variations in E minor follow, the first nine of which are rugged in character and greatly varied in all that Brahms adds above or below the theme itself. With the next two variations (10 and 11) tension grows less as they prepare

* There is a close parallel between the construction of this movement and that of Beethoven's Thirty-two Variations in C minor, for piano. (Ed.)

the way for the moving flute solo (Var. 12). Here in placid style and in a broader 3/2 time this instrument cloaks the theme in a disguise hard to penetrate. Variations 13, 14 and 15, still in 3/2 time, are set in the major key, each of them enveloped in a kind of mist. And when after being silent for ten variations the trombones re-enter at the fourteenth with solemn *pp* chords, remote and impressive, we certainly feel we are on the edge of some spirit-world far removed from reality. Then at the sixteenth variation the minor key and the 3/4 tempo return. At this, ten further variations are hammered out remorselessly on the full orchestra. The omission of trombones from the remaining five variations reduces the tension to a certain extent, though the underlying mood of agitation is but little relieved. Then a *Più allegro* coda of no less than fifty-nine bars follows. In a great outburst of wintry sounds, defiant in their opposition to the major key, Brahms takes the notes of the adapted *chaconne* theme and in partly canonic form drives them through a remarkable series of modulations which stand out like an epitome of all the many chromatic harmonies heard not only in this movement but throughout the symphony. The end is reached with no concessions to romance. So immense were Brahms's powers of invention that we can well imagine he could have continued his variations till kingdom come.

14
PETER ILYICH TCHAIKOVSKY
(1840–93)

Hans Keller

*

OUR age has produced the notion of the neurotic artist. Is he to blame or it? The critic of Tchaikovsky cannot help being interested in this question, because for many, Tchaikovsky is the neurotic artist *par excellence*. The so-called 'Pathetic' symphony (Tchaikovsky withdrew the title) fills the Albert Hall. Do eight thousand people assemble at a time in order to hear Tchaikovsky pitying himself? No doubt there are emotional complexes in the 'Pathetic' which, in life, can give rise to neurosis. (There are no neurotic complexes as such: it all depends on what you do with them.) No doubt some of the emotions which Tchaikovsky expresses with ruthless frankness can stimulate a self-pitying operation in many listeners, who may promptly project their feelings onto him. If they do so, *they* are behaving neurotically, not because they have been subjected to neurotic art but, on the contrary, because they are incapable of facing this particular piece of artistic reality – incapable, perhaps, of coping with strongly expressed emotion altogether. It may indeed be primarily because of the unambiguous clarity of his emotional expression that many sophisticated music lovers still underestimate Tchaikovsky as a symphonist, whereas his shallower masterpieces (the adjective is to be taken literally, not negatively) are fully appreciated – his ballet and 'salon' music which, like almost everything else he wrote, does of course offer us the unique flow of his melodic invention.

If we take into account the recently-published E flat major symphony (reconstructed and orchestrated by Semyon Bogatirev and first performed in this country in 1962), which certainly deserves an occasional hearing, there are altogether eight symphonies – none of them negligible:

No. 1 in G minor ('Winter Dreams'): 1866.
No. 2 in C minor (the 'Little Russian'): 1872.
No. 3 in D major (the 'Polish'): 1875.
No. 4 in F minor: 1877–78.
'Manfred' (in B minor): 1885.
No. 5 in E minor: 1888.
Symphony in E flat major: 1892.
No. 6 in B minor (the 'Pathetic'): 1893.

Tchaikovsky, it will be seen from this little table, is the only great symphonist who concentrated almost exclusively on the minor mode, which at his particular historical stage gives some indication of the emotional spheres he illuminated.

The first and weakest symphony was written at a time when the twenty-eight-year-old composer's spirit had been temporarily broken by a piece of music criticism: 'When I read this terrible judgement, I hardly knew what happened to me. I spent the entire day wandering aimlessly about the town repeating to myself "I am sterile, I am a nonentity, nothing will ever come of me, I have no talent".' We critics ought to have this quotation framed above our desks.

No work, then, was ever born under greater strain and pain, but its 'glaring deficiencies', as Tchaikovsky later called them, are almost exclusively formal, while its richness of invention, even where it is an embarrassment, foreshadows genius: it is the mediocre talents that tend to write 'good' early works, whereas latent genius urges towards over-invention, towards an abundance of tunes at the expense of the symphonic unity of the material. The scherzo of the symphony, incidentally, utilizes the scherzo of the posthumous piano sonata in C sharp minor.

'Little Russia' is (or was) the Ukraine, and the title 'Little Russian' (not the composer's own) for the Second symphony established itself in view of the Ukrainian folksongs used in the outer movements. The work exists in two versions; it would, to say the least, be instructive to have an occasional performance of the original, which Tchaikovsky rewrote and revised immediately after the first performance. The degree to which his symphonic thought had matured in the six years since the composition of the First is as striking as any later progress – and by 'symphonic thought' I do not only mean the development but, in the first place, the characterization of the material. Ukrainian tunes apart, it seems significant that whereas the First quotes from the past, the Second quotes, as it were, from the future: the basic thought of the second movement, *andantino marziale, quasi moderato*, was to grow, more than twenty years later, into the (not so called) march of the Sixth symphony's third movement.

As for the folk material itself, Tchaikovsky and Shostakovich are perhaps the only composers who invalidate Schoenberg's contention that folkloristic symphonies are a structural impossibility;* nor would Shostakovich's achievement in the much underrated Eleventh symphony have been possible without his assimilation of Tchaikovsky's folkloristic symphonism, the climax of which was to be reached in the finale of the Fourth symphony and the opening movement of the Fifth.

The Third, meanwhile, was his first symphony in the major mode – in fact the only one apart from the reconstructed E flat symphony of 1892. Begun on 5 June 1875, and finished – orchestration included – on 13 August, it shows Tchaikovsky's symphonism at its so far freest and most fluent. The five-movement scheme is no doubt influenced by Schumann's 'Rhenish' symphony, but the suite-like character of the symphony's formal layout, in which the central slow movement is surrounded by two scherzos, does not mean a regression of symphonic thought; on the contrary, with the introduction of dance rhythms into the material of every movement except for the slow one, Tchaikovsky

* Schoenberg's irrational obsession with structural 'possibilities' and 'impossibilities' is an inevitable late stage in the Germanic nightmare. Anything is possible in art, given the necessary imaginative genius. (Ed.)

widens the field of symphonic contrasts both within and between movements: the newly-charted territory was to be explored in depth in the Fourth. The dance characters themselves are, moreover, new or at least novel. The symphonic march, which we have noted in the previous works, now reappears in two movements, the first and the trio of the fourth: together with Schumann, Tchaikovsky here prepares for Mahler's – not to speak of Shostakovich's – highly fruitful marching compulsion. The *alla tedesca* of the second movement is less German than Beethoven's, and not Austrian like the symphonic *Ländler* of Mozart (string trio, clarinet quintet) or Bruckner, whose characteristic *Ländler* were to widen symphonic expression still further. But history's decisive step in the symphonic assimilation of the waltz was made by Tchaikovsky;* and although a Russian, he was one of the few great composers who could write a great waltz, in and outside the symphony. The Third symphony's waltz might perhaps be described as Polish: in its remoter background, one feels the pulse of the Mazurka. The nickname 'Polish' for this symphony, then, is not perhaps so superficial as is generally thought: the finale's polonaise – another symphonic conquest – is not the only Polish element in it, nor the subtlest.

The Fourth symphony is, in my opinion, the greatest – and its first movement is the most complex and innovatory of the four. The work dates from the most critical year of Tchaikovsky's life, the time of his abortive marriage, his attempted suicide, and his separation; but also the time of the beginning of his extraordinary friendship – they never met face to face – with his rich patroness, Nadezhda von Meck. To her inspiring influence we may owe much that is profoundest in this music, but we also owe her the profoundest artistic mistake which Tchaikovsky ever made: unfortunately, he acceded to her request to explain the 'programme' of the symphony. The result has proved noxious ever since. On the one hand, that is to say, almost every writer on the work has welcomed the opportunity of copying

* What about Berlioz? (Ed.)

this alleged programme instead of concerning himself with one of the most towering symphonic structures in our whole literature. On the other hand, Tchaikovsky's strictly private literary attempt has been thrown back at him by sundry anti-romantic, emotion-fearing neurotics. It would not, of course, be fair to include Alfred Einstein in this category, but the fact remains that he did allow himself to participate in our age's collective neurosis when he remarked, in his *Music in the Romantic Era*, that Tchaikovsky 'filled his purely instrumental music to the brim with a programme of feeling'. (As if feeling needed a programme – a typically contemporary fallacy!) Einstein proceeded to quote the entire 'programme' of the fourth symphony, concluding that 'Tchaikovsky let himself be led in his creative work by melodramatic and sentimental programmes such as this . . .' It is demonstrable that Einstein, a conscientious musicologist if ever there was one, was here inspired by a neurotic illusion – the simple fact being that Tchaikovsky never thought of this programme until he was asked for it after the event. In the same letter to Mme von Meck, moreover, he points out that by 'its very nature', instrumental music 'does not submit to the kind of interpretation' he himself has just given her. Finally, in a somewhat despairing P.S., he adds: 'This is the first time in my life that I have tried to transpose ideas and images into words [not, mind you, verbal ideas into music], and I have certainly not been successful.' Far more pertinent to the spirit of this symphony is a letter he wrote to Taneiev some years later: 'It's a thousand times pleasanter to write without a programme. In writing a programme symphony I feel as if I were a charlatan, cheating the public by giving it worthless paper instead of good coin.' And again, to Balakirev: 'This is the real truth: generally speaking I don't like *my own* programme music, as I've already told you. I feel infinitely freer in the sphere of pure symphony. . . .' The Fourth is a pure symphony.

Tchaikovsky's individual contribution to the development of symphonic thought was the discovery and integration of new and violent contrasts, of which the opening movement of the Fourth is perhaps the most outstanding example. It must indeed have been the puzzling wealth of its ideas, keys, developments and interrelations that was responsible for the somewhat mild success of the symphony at its first performance in Moscow. 'The public received your symphony very well,' Mme von Meck wrote kindly, 'especially the scherzo.' This is the kind of compliment which the Viennese historian and humorist Egon Friedell called 'manslaughter', i.e. unintentional killing.

So novel is the sonata structure of the first movement that in order to aid comprehensibility Tchaikovsky uses his introductory, basic motto theme for the purpose of defining his formal outlines: it recurs as a transition from the exposition to the development as well as at the climax of the development itself, and again by way of lead-back to the recapitulation. Eventually, it goes to form the transition to the coda, in which it then appears in imitation. One previous movement in the history of the *genre* had employed what was, fundamentally, the same method of formal clarification: the first movement of Beethoven's Fifth which, likewise, would have been too new to be clear without the strictest definition of its formal background.

Within and against this framework, Tchaikovsky develops his unprecedented contrasts. We know that the usual sonata procedure is based on the tension between the first subject in the tonic and the contrasting second subject in a related key* – prototypically the dominant – which pulls you away from the tonic. Now, Tchaikovsky not only increases the contrasts between the themes on the one hand and the keys on the other, but he develops a novel kind of contrast on an unprecedented scale – that *between* the thematic and the harmonic contrasts, which are therefore not allowed to coincide. Readers of the chapter on Mozart (see p. 84) may remember that, *pace* Beethoven's formal influence, I find Tchaikovsky's starting-point for this symphonic adventure in the music of his beloved Mozart, some of whose symphonic structures show a clear (if, to some, still confusing)

* But see Harold Truscott (p. 23) and Basil Lam (pp. 104 and 109). (Ed.)

interpenetration of formal, sonata elements. Tchaikovsky drives the process some decisive steps further afield. Thus the profoundly inspired theme of the second subject (see example) is not in the main contrasting key, i.e. the key of the *harmonic* second-subject stage, which is B major – but in A flat minor:

239

B major itself appears, to begin with, as a middle-section key of this second subject, which then returns to A flat minor; and by the time that proper second-subject *key*, B major, has established itself, the second-subject *theme* proper (i.e. the above example) is over and we have entered the enormously expanded closing section of the exposition. It is at this stage that the Mozartian models I have analysed spring immediately to mind. To sum up in simple terms: the *thematic* second subject precedes the *harmonic* second subject.

So much for the basic contrasts; now for their integration. Harmony first. B major, the key of the harmonic second subject, is really C flat major, i.e. the relative major of the thematic second subject's A flat minor (see example), which in its turn is the tonic minor of the home key's (F minor's) relative major. (If you read this sentence slowly it isn't really obscure.)

Thematically, the integration of contrasts is perhaps still more fascinating. The first subject, though sub-titled '*in movimento di valse*', keeps the waltz very much in the background, against which the theme sombrely develops; and it is only with the afore-quoted music example that the waltz emerges to the fore and, retrospectively, illuminates the background of the first subject, thereby establishing a unity with it. This gradual emergence of a rhythmic character, which at first remains latent if not indeed contradicted, is entirely a discovery of Tchaikovsky's. The procedure was to have a far-reaching if unobtrusive effect on the history of composition; even in so advanced a structure as the third movement of Schoenberg's violin concerto we can detect the ultimate influence of Tchaikovsky, i.e. when the basic march character refuses fully to establish itself until the first rondo return. (That Schoenberg is turning in his grave at this stage of an investigation is neither here nor there.)

The recapitulation of Tchaikovsky's first movement continues and elaborates the split between the thematic and the harmonic side of sonata form. Thus, the first subject reappears, not in the tonic, but in D minor, as indeed does the second (our example), thereby treating D minor as if it were the tonic – for in the formal background of Tchaikovsky's thoughts and one's own, first and second subject are expected to be recapitulated in the same tonality, which however would, of course, be the tonic. In the actual foreground structure, however, the tonic major only emerges in the middle of the thematic second subject (cf. example), in the same place where the chief contrasting key, the proper second-subject key, had appeared in the exposition, whose harmonic scheme is thus meticulously retained despite the simultaneous, lavish variation of harmonic events. At the same time, the home mode, F *minor*, is saved up for the coda – with the result that the climactic restatement of the first subject with which the movement ends assumes the tense significance of a first complete recapitulation: long as the movement is, one is almost shocked that it is already over.

The ternary structures of the middle movements are more immediately comprehensible, though no less inventive and characteristic. But the structure of the finale is less simple than it has seemed to all commentators I have come across. In the textbooks, that is to say, the movement is described as 'a set of free variations'* upon a Russian folk-song, *In the fields there stood a birch-tree*. In reality, it is not a variation movement at all, but another of Tchaikovsky's

* I think we should always be on our guard against 'sets of free variations'. Formally and structurally, the concept is meaningless, and Beethoven's 'Choral Fantasy' as well as the finale of the Ninth symphony is amongst its chief victims: both structures are composed against the background of sonata form.

intriguing sonata structures. As in the first movement, a decisive contrasting stage – this time the thematic second subject itself – is anticipated by the middle section of the preceding stage – here the first subject: after a mere eight bars of the first subject in the tonic major, which has been the key of the scherzo, Tchaikovsky jumps upon us with the folk-song in A minor which, after the first subject has been resumed, turns into the second subject. It is a tune which confines itself to the first five notes of the diatonic scale, like the 'Joy' theme from Beethoven's Ninth: one was chosen, the other composed as the simplest possible thematic basis. At the second-subject stage proper, it is treated as an *ostinato* climax in B flat (the subdominant) minor, the key of the slow movement; and in the recapitulation it returns, modified, in D minor, the tonic's relative minor – as did the thematic recapitulation of the first movement. The symphony's motto theme finally comes back between the end of the recapitulation and the coda as, again, it did in the opening movement. This is indeed an ideal spot for such a free insertion: in the background, there hovers the idea of the cadenza which, in the first movement of a classical concerto, unfolds precisely at this formal juncture.

The 'Manfred' symphony may not derive much structural benefit from its Byronic programme, but it still does not, as Tchaikovsky himself put it, 'cheat the public by giving it worthless paper'. In much of it, especially in the first movement, the patient listener will find a great deal of Tchaikovsky's originality as we have described it, and there is no doubt that it would be far more often performed if it were not sixty-five minutes long, and so difficult to play. Uneven it is, certainly, but there are stretches that are as good as, if not better than, anything Tchaikovsky's genius produced. Paradoxically, therefore, his own reactions to the work, seemingly contradictory, do not cancel each other out: 'I value it very highly . . . I may be wrong but it seems to me the best of my compositions' (to his publisher, 1885–6);* 'Without any wish to make a show of modesty, I can say

* The translations are by Gerald Abraham.

that this production is abominable and that I deeply loathe it, with the single exception of the first movement' (to the Grand Duke Constantine, 1888).

The Fifth symphony, like the Sixth, is so popular that one feels slightly embarrassed to write about it; it is as if one were invited to write an essay recommending the 'Blue Danube' waltz. Yet, at the present stage in the history of music, one finds oneself in the paradoxical position of having to defend these works against their popularity.

March and motto are now rolled into one, and no contrast is more surprising, and at the same time more natural, than the reappearance of the motto, in triple time, in the coda of the inspired waltz* which, as third movement, takes the place of the scherzo, and which itself establishes a more drastic contrast to the slow movement and the finale than any more traditional scherzo form could have hoped to achieve within a similar context. The symphony, Tchaikovsky hoped, would 'prove not only to others but also to myself that I am not yet played out as a composer'. The integration of folkloristic material shows him, in fact, at the height of his powers as a composer in the more literal sense: on purely internal evidence, it would be impossible to discover that the opening movement's first subject is not Tchaikovsky's own invention, but is built on a fragment from a serious Polish love song – so eminently symphonic does this basic material seem in the light of one's wisdom after the event. At the same time, the total structure of the Fifth is not as consistently impressive as that of the Fourth; at times, as Tchaikovsky observed, 'the organic sequence fails, and a skilful join has to be made . . . I cannot complain of lack of inventive power, but I have always suffered from want of skill in the management of form'. Not – I hope to have shown – always. Besides, in one respect, the symphony may be the most consistently outstanding of them all: the orchestration offers original sounds at every change of texture. If this is not generally recognized, it is only because all

* Clearly inspired by a similar stroke in Berlioz's *Symphonie fantastique*. (Ed.)

these sonorities seem as natural and necessary as the hills.

The reconstruction of the E flat major symphony of 1892, a ten-year job, was finished and published in 1961 (State Music Publishers, Moscow). In the score, the editor gives a detailed description of what he has done. Two main points emerge. (1) In 1893, much of the music of the unfinished work was turned into the third piano concerto. (2) The scherzo in the reconstruction is the *Scherzo-Fantaisie* from the Eighteen Piano Pieces, Op. 72, which did in fact start life as an orchestral piece: the editor, following Modest Tchaikovsky, has many good reasons for thinking that it was intended for this symphony. In any case, the work is intermittently impressive and characteristic, the editor's orchestration both faithful and skilful: let us not altogether forget it.

The *Pathétique*? No other work has survived so many intellectual burials. Gerald Abraham has shown that the draft programme for the Sixth symphony was a note which Tchaikovsky scribbled on a scrap of manuscript paper, probably in 1892: 'The ultimate essence of the plan of the symphony is LIFE. First part – all impulsive, confidence, thirst for activity. Must be short. (Finale DEATH – result of collapse.) Second part love: third disappointments; fourth ends dying away (also short)'. The note is trite enough if you think that these words were written down in order to produce music; in reality, of course, they were a brief verbal memorandum of the mighty musical structure that was gaining shape in the composer's mind. 'Finale DEATH' – 'so what?' we might ask before we realize the musical idea behind the seeming platitude. 'There will be many formal innovations in this symphony', Tchaikovsky later wrote to his nephew, 'among which the finale will not be a noisy *allegro* but, on the contrary, a most long-drawn *adagio*.'

The symphonic world never was the same again after this Sixth symphony had been performed. We speak of Tchaikovsky's over-statements, but had full-scale sonata form ever been so drastically, and so consistently understated, compressed, as in the first movement, with its violent shortening of the second subject and its grippingly compact coda? Is there a more original transformation of scherzo function, a more breath-taking contrast, than that ultimate metamorphosis of the Tchaikovsky waltz that is the second movement in 5/4 time? And then there is another scherzo, in fact the scherzo proper: the two scherzos, the inner dance movements of the Third symphony, have now moved right into the middle; and the march of the Second symphony appears in the second scherzo, i.e. the third movement, by way of that gradual emergence of unexpected rhythmic character which (as we said apropos of the first-movement waltz in the Fourth) was one of Tchaikovsky's most individual discoveries. Where would Mahler's and indeed Schoenberg's symphonic structures be without Tchaikovsky's 'many formal innovations' – above all, without that *Adagio* finale itself, which was to cast an illuminating shadow over the whole future history of the form? This time, Tchaikovsky *had* 'played himself out': a week after the symphony's first performance he was dead.

15
ANTONIN DVOŘÁK
(1841–1904)

Julius Harrison

*

NOW that LP records of the four early symphonies are available we are the better able to appreciate the significant development of Dvořák's symphonic style from its early days (1865) to its final efflorescence in the immensely popular 9th symphony in E minor – 'From the New World'. With that development came, perhaps a little strangely or at least unexpectedly, a marked simplification in the use of the orchestral instruments, while at the same time the music itself grew in lyrical intensity to such a point of rich fulfilment in the five well-known symphonies that there can be no denying that Dvořák was one of the greatest and most inspired of the later nineteenth century composers. His genius, his unassailable reputation as a magnificently endowed composer, are, in every aspect of his art, a living portrait in sound of the man himself: a man with a noble nature, deeply sincere and unaffected. 'In spite of the fact that I have moved about in the great world of music', he once wrote, 'I shall remain what I have always been – a simple Czech musician.'

His art was primarily a homely one, engaging, lovable; only a misanthrope could resist it; the music springs from a peasant mind, God-fearing and devoid of all guile. He was the natural successor to Schubert. Like him he spun melody after melody from the silk of his fertile imagination. And outside his working hours his pleasures were the simple ones – again like Schubert. When a village band, to do him honour, once painstakingly rehearsed operatic music for him to hear, he would have none of it. 'No', he said, 'play me some village music; that's what I like.' His main pastimes or diversions were visits to railway stations and yards 'loco-spotting'; looking after his pouter and fantail pigeons at his summer retreat at Vysoká or playing cards or skittles with boyish delight.

For a while he served in his father's butcher's shop, though his musical studies were by no means neglected. And then eventual release from such rough manual drudgery came in 1857 when he left coal-mining Zlonice to become an organ student in Prague, to join Komzak's band there, and, as viola player, to be later accepted in the orchestra of the Provisional National Theatre where only Czech operas were performed. In that practical school he picked up his knowledge of orchestration with its theatrical panache, its stirring rhythms *à la* 'Bartered Bride'; and, through visits to the German Opera House, much else besides. As a result of going there he came under the spell of Wagner, and it took him some years to expunge this influence from his own compositions. In the early symphonies we hear him employing sonorous brass chords much in the style of *Das Rheingold* or *Die Walküre*. Or again, flirting innocently with the Venusberg type of sounds in *Tannhäuser*, or with those static woodwind harmonies which inevitably remind us of Holy St Elizabeth and Wolfram's 'Star of eve' soliloquy. But by 1875 – the year of the F major symphony – he had left most of these influences behind him. Like Smetana, he was now the ardent Czech patriot in all he wrote. Yet the older he became the more he clung to the classical symphonic forms, even though in the actual music the vehement spirit of Bohemian village life lurked in countless fiery phrases springing from his vivid imagination. He was not, as Sibelius was to be, concerned to challenge the traditional German forms accepted throughout Europe as the *sine qua non* of a symphony's architectonics. He reached out to Beethoven and Brahms, yet could still preserve his Czech identity.

Since Dvořák so often altered the opus numbers of his compositions, sensitive as he was to the wide stylistic gulf that lay between his early and more mature works, and because his chief publisher Simrock did not issue the 1875 F major symphony till 1888, wrongly calling it No. 3, much confusion has arisen in the matter of numbering the nine symphonies and the chronological order in which they were written; until more recent years little was known about the four early symphonies. The list below shows the numbering which has now been generally accepted, together with the old numbers in brackets:

No. 1	C minor	1865		
No. 2	B flat major	1865		
No. 3	E flat major	1873		
No. 4	D minor	1874		
No. 5	F major	1875	(No. 3)	op. 76
No. 6	D major	1880	(No. 1)	op. 60
No. 7	D minor	1885	(No. 2)	op. 70
No. 8	G major	1889	(No. 4)	op. 88
No. 9	E minor	1893	(No. 5)	op. 95

Dvořák burnt many of his earlier compositions, a fate the four early symphonies managed to escape, two of them by sheer luck. The C minor symphony was saved only because the score went astray, later (1882) to be found in a Leipzig second-hand bookshop, and then turning up once again in 1924 in the collection of Dr Rudolf Dvořák (no relation). More curiously still, the B flat symphony survived because Dvořák could not afford to have the score bound! This task his National Theatre colleague Mořic Anger undertook to superintend; and since Dvořák could not find the money, Anger held on to the score, fearing its composer might destroy it – which he would certainly have done after Adolph Čech had given a performance of the symphony in 1887. But for the E flat symphony Dvořák showed more regard, even affection. The Prague Philharmonic Orchestra, Smetana conducting, performed it in the Žofin Island Hall on 30 March 1874. A year later it gained the composer an Austrian State prize, Brahms being one of the judges. The D minor symphony (not to be confused with the Op. 70 symphony) had its scherzo played, again under Smetana, in 1874; and in its entirety was conducted by Dvořák himself in 1892.

The C minor symphony, 'The Bells of Zlonice', may well

have been Dvořák's first orchestral work, composed only eight years after he left Zlonice for Prague in 1857 as a raw student, untutored and unskilled. Not only in this fact but in all other respects, it is a remarkable example of its composer's natural genius; an almost incredible outpouring of mature – and less mature – ideas worthy of many performances in the concert-hall. Its decorative sub-title is not found on the full score, though Dvořák had indeed toyed with the idea of adding it. A brief but impressive *Maestoso* introduction leads to an *Allegro* first movement that is serious and even gloomy in mood; portentous tribulation-music written in the grand classical manner. The second movement, an *Adagio molto* in A flat, is truly Beethovenish in its many long *cantilena* passages of considerable beauty. While it perhaps outstays its welcome, yet at the same time relief is obtained through some inspired moments of fantasy such as the fanfare for horns and trumpets preceding the brief but lovely recapitulation. By turns serious and lively, the *Allegretto* third movement reaches a brilliant peroration after the flow of the music has been interrupted several times, much in the manner of the scherzo in Beethoven's Fifth. The *Allegro animato* finale, while more formal in ideas than the rest of the symphony, never falters in its progress towards climaxes of irresistible C major splendour, in whose culminating, fanfaring coda sections we can indeed imagine those bells of Zlonice vividly heard again by Dvořák as a fond and thrilling memory – now magnified into a full symphonic peal such as would ennoble the tower of some great cathedral. The concert-going public should be given every opportunity of hearing this remarkable first symphony. Its *première* took place in Brno in 1936.

As a complete work the B flat symphony is not the equal of No. 1, though there are many beautiful and thrilling moments in it. In its first two movements it is a lovelorn work, sumptuous in scoring, rich in overcharged emotions welling from the heart of a composer who at that time had been rejected by the unattainable lady he adored. (She was

Josefa Čermáková, whose sister Anna he eventually married in 1873.) That Dvořák at this juncture turned to Wagnerian sounds is scarcely surprising. The chord of the major 9th – that luscious chord Venus and her troop made so much their own in *Tannhäuser* – dominates various climaxes in the symphony's first and last movements. The *Poco adagio* movement contains much lovely music of a solemn character, though in time it becomes tiresome through the reiteration of small woebegone phrases. Then in the third movement, *Allegro con brio*, the mood suddenly changes to one of boundless joy, made all the more attractive when the music breaks into a heart-easing melody as lovely as anything Dvořák ever wrote. Well could he here have sung with Chaucer:

> Sin I fro love escaped am so fat,
> I never thenk to ben in his prison lene;
> Sin I am free, I counte him not a bene.

The finale, *Allegro con fuoco*, reverts considerably to the Wagner mood and is a truly exciting movement, noisy with Tannhäuserish trombones as it works up to climaxes of immense power often created out of indifferent thematic material.

There are only three movements in the E flat symphony: a flowing, yet impetuous and magnificently constructed *Allegro moderato*; an *Adagio molto* (*Tempo di marcia*) set mostly in C sharp minor and D flat major (so is the *Largo* in the 'New World' symphony), and a one-in-the-bar *Allegro vivace* finale which, for sheer excitement and ceaseless joyous drive, has a certain kinship with Schubert's in his great C major symphony. Dvořák's orchestration throughout reaches its most ornate expression. The strings are subdivided more elaborately than they were ever to be again; various instruments many times double notes played by others in a score overflowing with sumptuous Wagnerian touches; a cor anglais and harp are added to the normal orchestra. It is as if Dvořák were having one of his last flings in alien (i.e. Wagnerian) sounds before gradually settling down via his next symphony to a more authentic Czech style. After this E flat symphony we begin to note the marked simplification of orchestral expression already mentioned. If the funereal slow movement tends to drag its length (fifteen minutes), there is yet so much in it to reprove our impatience, for it aspires to the nobility of Beethoven's 'Eroica'. In its D flat middle section (110 bars) Dvořák builds up a truly magnificent climax out of the most naïvely simple phrases. This symphony should be performed again and again. No wonder Dvořák loved it; no wonder Brahms thought it worthy of that Austrian State prize.

When Dvořák composed the D minor symphony of 1874 the flood of Wagnerian sounds had not altogether subsided. But the green fields of his native Bohemia were once again beginning to show quite plainly above what remained of the swirling waters. This last of the four early symphonies can be regarded as a transitional work that caught Dvořák in two opposing moods. For in the first *Allegro* movement there is a certain epic grandeur coupled to rather empty rhetoric of the 'diminished seventh' order, as if the composer were doing his utmost to exorcize some malign spirit from the body of his music. The next movement, *Andante e molto cantabile*, consists of a series of variations in B flat founded on an extended theme of seventeen bars, foreshadowing much of what was to come in the slow movements of the later symphonies. Here Dvořák reveals his most lyrical self and his genius for clothing a theme in richly embroidered apparel. The *Allegro feroce* scherzo is an intriguing medley of fiery Czech rhythms founded on a catchy tune that could well take its place in some imagined operatic ensemble. The trio, again operatic in style, is a march-patrol swelling to an exciting climax and then dying away before both scherzo and trio are briefly recapitulated. Irresistibly the trio music, with its many trills, reminds us of the entrance of the Tailors' Guild in *Die Meistersinger*. The *Allegro con brio* finale is far less attractive, so dependent is it on a rūm-tūm, rūm-ti-tūm rhythmical figure irritatingly repetitive; not even interpolations by a flowing (but commonplace) theme can make its many returns any more endurable. Was it perhaps this finale that made Dvořák overhaul his symphonic machinery, giving it, in the F major symphony of 1875, a new lease of life still more national in character and as yet uninfluenced by Brahms?

With the F major symphony Dvořák began the sequence of five in which endless melodies both lyrical and dramatic were to enrich the music as never before. Even if in the two most important symphonies, the D major and D minor, he turned towards the type of melody Brahms cultivated, he still could not avoid expressing his own rugged self. Dvořák's biographer Šourek has likened the first movement of the F major to 'the voice of the rustling woods, the song of the birds, the fragrance of the fields; the strong breath of nature rejoicing and a sense of mortal well-being'. Its opening theme, which Tovey calls 'the lightest symphonic opening since Beethoven's Pastoral Symphony', comes from the clarinets, simple yet rhythmically alive, founded for eight bars on no more than the triad notes of F and supported by a drone bass on horns, violas and 'cellos. Not till the ninth and tenth bars is there a cadential change of harmonies. This innocent theme, here quoted in full, is truly characteristic of Dvořák's simplified style; something even unmusical Charles Lamb would have approved when, long years before Dvořák's time, he wrote to Coleridge: 'Simplicity springs spontaneous from the heart and carries into daylight its own modest buds and genuine, sweet and clear flowers of expression.'

240

Even the frequent *ff* outbursts which it and the two other main themes subsequently engender carry with them Šourek's 'sense of mortal well-being' and none of that feeling for epic tragedy or grandiose display encountered *passim* in the early symphonies. While the trumpets and trombones still apply their brassy weight in moments of climax, there is now a minimum of 'diminished sevenths' melodrama. But to quote the other themes is not possible here; their style is almost as innocent as the opening theme itself. From the movement's final *ff* climax – two themes are combined here – to the placid, contented homespun *pp* which marks the end, Dvořák writes a coda as lovely in its echoes of the clarinets' theme as it is in its serene orchestral expression.

The second movement with its 'cello melody – soon transferred to violins and then to woodwind – certainly springs spontaneous from Dvořák's heart. Set in the key of A minor, and in A major for its middle section, its moods of gentle melancholy are somewhat akin to that form of composition, the *Dumka*, to which Dvořák so often turned in his chamber music.* In place of the solemn, even portentous moods characterizing the slow movements of the four early symphonies, he now contents himself with the simplest melodic phrases harmonized with a like simplicity.

241

Trumpets and trombones mark two brief moments of climax with no more than fourteen bars of loud chords. Otherwise they have next to nothing to say. The string and woodwind writing throughout is equally simple while still lacking nothing in aptness or sad beauty. Significant, too, is the continual use of figurative demisemiquavers derived from

* *Dumka* (plural *Dumky*, 'lament') comes from Ukrainian sources. Its equivalent in Roumania is the *Doina*, possibly derived from the Sanscrit *d'haina*. In Tudor times composers sometimes indulged in a melancholy tune called the *Dumpe*.

the first four notes of Ex. 241. These notes give rhythmical life to the whole movement after their introduction as running accompaniment to the woodwind when those instruments take up the 'cello melody. After the movement has ended, Dvořák toys with a few afterthoughts (sixteen bars) founded on Ex. 241, in this way modulating a little timidly from the A minor key to that of the scherzo movement in B flat major.

In this invigorating scherzo and its equally attractive trio Dvořák scales classical heights in a way hitherto unapproached. But for the absence of an exciting coda to round off the whole movement, it is a foretaste of racier Slavonic Dances Dvořák was to make so popular three years later. Classical in its form – scherzo, trio and *da capo* scherzo – it continually breaks down the monotony of endless four-bar phrases (common to so much music) by interposing odd lengths of three, five and six bars, thus creating the fluidity of rhythm characteristic of all great symphonic music of the eighteenth and nineteenth centuries. While the opening theme only can be given here, the many rapid and Schumannesque exchanges of brief phrases

242

between woodwind and strings in the D flat trio have an equal importance. In due course these precipitate a short-lived, brassy climax that balances to a nicety similar outbursts in the scherzo section.

Dvořák plunges dramatically into the finale with a striking theme through which he not only remembers the A minor key of the second movement but also the need to consolidate the whole structure of the symphony with music set in a broader time-signature than had hitherto been employed.

243

His ultimate goal is of course F major. But in this remarkable example of what is called 'progressive tonality' that goal is reached only after a veritable battle royal with the determined A minor key – an onslaught here parried if not yet overcome after 54 bars of strenuous struggle. Then from this F major version of Ex. 243 prolonged development follows, especially after a suave 'second subject' in D flat major where the general tension is replaced by flowing minims and crotchets. But the strenuous climaxes soon return and eventually Dvořák adroitly steers the music back to the A minor theme, now reintroduced by the oboe and then followed, as a complete surprise, by the *bass clarinet* in its only contribution to the symphony: seven bars!* From now on much of the music is either restated or varied both in tonality and orchestration as with sure steps it approaches its final F major victory. Here Dvořák remembers snatches of the symphony's original clarinets' theme (Ex. 240); first of all quietly on flute, then clarinet, and finally on a trombone in a veritable full-orchestra cornucopia of F major sounds made all the more brilliant by fanfaring trumpets competing with the trombone for mastery.

* A capricious feature in other later Dvořák symphonies is the odd phrase of a few bars given to instruments otherwise unemployed: e.g. piccolo and cor anglais (first movement in G major symphony); piccolo (first movement of the 'New World'); and the single cymbal stroke in the same symphony's finale. At a concert under Hans Richter many years ago the cymbalist missed his cue. Richter never forgave him. Years afterwards, once more rehearsing the symphony, he hopefully enquired of his leader: 'Dat Becken man, is he dead?'

By 1880 Dvořák was well acquainted with Brahms's C minor and D major symphonies, and to the latter particularly he was drawn with an enthusiasm which surely must have spurred him on to the creation of his own D major masterpiece. All the same, too great stress should not be laid on his indebtedness to Brahms. Such thematic similarities and other like features as exist belong to the outer movements alone and intermittently at that. Both his and Brahms's symphony have two things in common: a 3/4 first movement and a 2/2 finale – coincidences (if such they are) which in no way detract from the sheer magnitude of Dvořák's work. He was no conscious plagiarist, however much he came to learn from his great contemporary about matters of musical restraint and discipline. And so the formal classical design of this symphony is far in advance of everything else he had so far accomplished. Themes, counterpoints, harmonies, rhythms are all of a piece. The music, never lacking variety, is set down more convincingly than ever before; music transcending its own national style, and, like Beethoven's, speaking with a universal tongue. Restraint, too, is shown in the exclusion of trombones and bass tuba from the two middle movements, making their impact on the finale tremendously exciting and cumulative. That Dvořák could dispense with them in the *Furiant* movement* – and without loss to the music – speaks volumes for this new spirit of restraint; for this Bohemian dance, as we know from Smetana's *Bartered Bride*, is indeed a lively cross-rhythmed affair. Here is the quiet start of the first movement:

244

* Its name has no etymological connexion with our 'fury'.

From this innocent theme, shared by various instruments above and below syncopated chords on violas and horns – a theme which (in Tovey's words) 'presents us with those intimations of immortality that make the child sublime' – Dvořák plunges at once into a mood of vehement self-expression peculiarly his own. Briefly then, it should be noted how within a few bars the sudden rise in the music's temperature sets the pattern of the whole movement, and how this childlike opening soon grows into an endless flow of other important and ancillary themes which together, impetuously and otherwise, create a symphonic first movement far more continuous and richer in contrasts than anything Dvořák had so far attempted.* Its design is truly classical. With its flowing second subject ('cellos and horns), its oboe pendant, and the later development section whose mysterious start is as impressive as any similar passage to be found in Beethoven, Schubert or Brahms, the movement proceeds unfalteringly to its stirring climaxes and final coda.

Nor is the *Adagio* movement any less moving, even though it seldom loses contact with the four-note phrase that from the first bar serves to introduce the main theme and then becomes naïvely repetitive.

245

The four bars foreshadowing this theme carry with them a faint reminder of the opening of the slow movement in Beethoven's Ninth symphony. Indeed, this whole long movement, its many lyrical passages shared by

* Compare the static F major mood of Ex. 240 (Fifth Symphony, p. 360) with the adventurous Ex. 244. In the first forty-four bars developed from Ex. 240 only one 'modulation' occurs; in forty-eight bars similarly in Ex. 244 there are eleven.

strings, woodwind and horns, recalls the same influence, even though Dvořák with his genius for rich harmony surrenders to moods more sentimental than any Beethoven would ever have permitted himself. There are several passages in this movement which provoke sudden but short-lived *ff* climaxes, the most notable of which (in B flat minor, nine bars only, founded on (*a*) in Ex. 245) soars starwards in a rapturous outburst not a whit inferior to what Beethoven might have written at this point. But it is not possible to describe here the many lovely orchestral touches abounding throughout this movement; each instrument, 'cellos especially, makes its contribution to a slowly revolving kaleidoscope of variegated colours. The quiet closing bars are as exquisite as anything Dvořák ever wrote.

For a scherzo Dvořák turns to the impetuous cross-rhythms of the *Furiant*, out of which he constructs a third movement as classical in form as any example by the great masters.

246

The music of the first section – the scherzo with its *da capo* repeat – is subjected to sudden, even violent changes in nuance: every few bars *fortissimo* and *pianissimo* jostle each other in an exciting interchange of brief phrases between wind instruments and strings. Then the slower-paced trio, largely dependent on the bracketed figure shown in (*a*) of Ex. 246, comes as a tranquil change from the *Furiant* proper. For seventy-six bars the music clings lovingly to its D major tonality before a gradual return is made to the scherzo section. The two piccolo phrases, each no more than four bars long and heard near the start of the trio, breathe a tender thought or two expressive enough to abash any fifer prone to indulge in ear-splitting excesses.

More extended in its range and imaginative continuity of thought than any other movement by Dvořák, the finale crowns the whole symphony in eloquent majesty. Its hushed opening, for a few bars recalling Brahms's D major finale, promises from its very first notes a grand *dénouement*. While the general design of this great symphony contains no cyclic recurrence of themes from other movements (such as characterize the 'New World' symphony so fortuitously) it is worth noting how the bracketed notes shown in (*a*) of Ex. 247 have a definite relationship to those shown similarly in Ex. 244.

247

In this way Dvořák intuitively consolidates the whole structure of the work. From its happy start this finale proceeds untroubled throughout its great length, its many themes and dramatic climaxes unfolding superbly in their approach to the *Presto* coda, which, in *fugato* style (Ex. 247(*b*)), transforms the hushed opening theme into something as strong and malleable as steel.

Prior to the year of his second D minor symphony Dvořák had already visited England twice and had been given a great welcome. The natural outcome was a request by the London Philharmonic Society that he should compose a new symphony for them. He responded with enthusiasm. 'Now I am occupied with my new symphony (for London)' he wrote to Antonin Rus, '. . . which must be such again as to make a stir in the world, and God grant that it may'. This symphony, the finest of them all, was conducted by Dvořák in St James's Hall on 22 April 1885. Its success was immediate and its musical worth soon recognized everywhere.

The work came at a time when Dvořák's mind was sorely troubled by many conflicting emotions. The loss of his mother affected him deeply, while at the same time his conscience was in a state of bitter opposition to his many importuning friends who thought that his compositions should now accord more closely with German ideals and principles. So grave, even austere, are its themes, its dark harmonies and its Brahmsian concept of classical form, that it holds a unique place in Dvořák's symphonic output. The haunting lyricism in the second movement affirms that 'the beautiful things in it are not mere *delights*, they are *depths*';* there are infectious rhythms together with a sun-warmed landscape in the scherzo, and then a return to reality in the finale. The opening theme of this symphony with its modal flattened sevenths (x) was Dvořák's truly Bohemian answer to these critics who strove to distract him from the immutable spirit of his art.

248

Here is Dvořák at his finest in a unisonal, dark-hued theme for violas and 'cellos fraught with immense possibilities (of which he took full advantage) in the two features marked (*a*) and (*b*). And if, in pursuit of tragic thoughts, chords of the diminished seventh are once again freely used, they now return in a new guise, shorn of the splashy rhetoric heard in the earlier D minor symphony. (Sibelius likewise made a new thing of this much abused yet dramatic

* W. R. Lethaby on the work of William Morris.

chord). Many *fortissimo* climaxes impetuously breaking out into sudden bursts, often of brief duration, help to galvanize the whole movement as it drives forward to its *accelerando* culmination before the closing bars echo the opening theme in mysterious, disjunct *pianissimo* phrases spread over many instruments. In actual notes (though not otherwise) the beautiful second theme closely resembles the 'cello solo in the slow movement of Brahms's B flat piano concerto.

249

In the first part of the *Poco adagio* movement, which opens with a fine and simple clarinet melody, Dvořák clings to the key of F major for no less than thirty-nine bars.

250

Into them he compresses a succession of themes as varied as one could wish. The clarinet tune (Ex. 250) is soon followed by another on flutes and oboes: something far more emotional by reason of its chromatic harmonies than we might have thought possible so near the start of the movement. Significantly it tells us that however placid the mood may be from time to time, we can expect little slackening of the general tension throughout. A melancholy theme, played in octaves by violins and 'cellos, is followed by horn and clarinet, which, as Tovey remarks, 'play the parts of a rustic Tristan and Isolde to a crowd of orchestral witnesses'. Extended development follows, the music modulating through many agitated passages with the trombones underlining its more tragic phrases. Then the F major themes return, but in different order. 'Cellos are given the one first heard on flutes and oboes; violas now join the violins in the sorrowful theme. A climax follows. After it, the oboe plaintively recalls a few bars of the clarinet melody, and then the movement fades away in a serene coda. That Dvořák in this recapitulation could change the order of his three main themes without hurt to the general design of the movement is a feature exceptional enough to be worth mentioning here.

The scherzo, its 6/4 time-signature broadly determining a movement more serious than playful, contains some of Dvořák's finest music, all growing out of a theme thoroughly native in its style.

251 

Impetuous in its syncopated rhythms, replete with *sforzando* effects, this compelling theme sticks in the memory long after the symphony has ended. Finely-wrought counterpoints add rich textures to a score which dispenses with trombones and yet maintains its forceful character. The scherzo's more tranquil middle section (Dvořák rightly omitted the word trio here) comes like a breath of country air, twittering with birdsongs; the whole an early-morning pastoral more gently lyrical and felicitous in its orchestration than can be heard anywhere else in the nine symphonies. After the return of the scherzo music, a coda, which contains a poignant outburst from the violas, rounds off the movement in tumultuous style.

No less inspired is the whole of the finale; and no less serious and impressive are the various moods unfolding from its tense opening theme. (The bracketed notes shown here occur later and often.)

252 

At the thirty-sixth bar flutes and oboes hint at a second and very important theme soon to be used with dramatic emphasis. But so profuse and inspired are Dvořák's inventive powers from this point onwards that each theme seems to contend for the right to be considered the chief motivating force of all that happens. Integration is complete: it is a case of *primus inter pares*. Dvořák now scales the heights of Parnassus as he was never to do again. At the right moment a beautiful third theme (in A) steals in on 'cellos to bring a touch of warm solace to music mostly given over to tragedy. Finally, at the end of a great classical coda, everything tragic and sublime is resolved in a succession of D major chords as exultant as they are inevitable.

With the G major symphony Dvořák's style underwent something of a change. Melodies come and go in more rhapsodic continuity, less concerned with problems of inner contrapuntal development. Their innate charm is matched by rich yet simple harmonies mostly of a diatonic character, springing naturally from the primary triads of the key in which the music happens to be at the moment. The spirit of fanciful improvisation hovers around, guiding Dvořák into byways far from that classical road never once deserted by the D minor symphony.

When this symphony was being composed Dvořák was overwhelmed with one commission after another. That its score is the plainest of them all makes one wonder whether sheer necessity proved to be the mother of its composer's invention, for it reveals (not to its detriment) an almost conscious effort to avoid all those elaborate instrumental touches a more complex symphony would have demanded. Published in London by Novello after Dvořák had had many a tussle with his Berlin publisher Fritz Simrock – 'I won't allow myself to be done down by him any more' – the symphony was slow to be acclaimed abroad. Now it has become a standard and much-loved work in the repertory of all orchestras.

Stepping out with a calm G minor theme for 'cellos,

clarinets and horns in unison – a theme rich enough to find its way *en route* through D flat major harmony – Dvořák follows it at once with two others in G *major*: one (*a*) first heard on the flute; the other (*b*) on divided violas and 'cellos. Both play important parts.

253

The G minor theme later returns twice; the second time on trumpets in a tense and dramatic passage that seems defiantly to dispute the right of the symphony to consider itself in the major key. This conflict or alternation between major and minor is indeed a typically Slavonic characteristic persisting throughout most of the symphony. What (for want of a better term) we can call the 'second subject' (Ex. 254) is set in B minor, returning later in G minor. Only Dvořák or Schubert could have imagined so delightfully naïve a tune.

254

The brilliant G major coda ending this movement leaves no doubt that however many and determined these minor-key excursions may be, nothing can deflect the work from its underlying mood of cheerful outdoor exuberance.

When Dvořák's authoritative biographer Šourek imagined the second movement as reminiscent of 'an ancient castle in various aspects, proud and melancholy, and, as second subject, a noble knight singing a love-song to his lady', he was toying with the fanciful dreams of folk-lore; thinking of Smetana maybe, and of all those legendary figures whose deeds are recorded in *Ma Vlast* or in his country's history. Even so, this beautiful elegiac movement, alternating dramatically between the minor and major keys of C and in this and other ways recalling the *Marcia funèbre* in Beethoven's 'Eroica', needs no such pictorial aids to our understanding of it. It is absolute music, plumbing the depths of sorrowful expression and then aspiring later to triumphant C major allusions to this opening theme.

255

Later, a few improvisatory bars for solo violin, followed by a remarkable passage for the whole orchestra in which scale passages predominate, bear further witness to the fact that this movement is the outcome of some deeply-felt experience. Was it Beethoven-inspired? Or was it Dvořák's unerring intuition guiding him to profound thoughts providing more substance for a symphony otherwise light in weight? Trombones and tuba are silent throughout, as they are in the next movement, the gentle *Allegretto grazioso* – which can scarcely be called a scherzo.

Except for its vivacious coda in 2/4 time, this 3/8 movement lilts happily along almost like a slow waltz. Once more the music is in G minor and major, but in contrasted A–B–A sections, the middle section and the final coda being in the major. No feeling of conflict between the keys remains; coexistence rather, just as if it had been finally agreed that a policy of live and let live was the only Q.E.D. to the symphony's strangely diverse modal problems. The melodies in both sections are transparently clear, those in the middle section taking on new shapes in the *molto vivace* coda. Dvořák's invention is here so delightfully naïve that I can only echo Hermann Kretzschmar's words that the whole movement 'looks at one with the eyes of a child'.

By way of introduction to the finale, Dvořák, showing the same childlike pleasure, summons us to the fair-ground, bright trumpets telling us to hurry along and not miss the fun.

256 

This finale is mainly a series of G major variations in many different moods on a beautiful 'cello theme (one closely akin to Ex. 253 (*a*) of the first movement):

257 

The trumpet flourish also plays a striking part later on, adding further gusto to music alternating between rhythmical liveliness on the full orchestra and quiet passages given to solo woodwind instruments or to strings alone. The peace of the countryside is never far distant from the din of the merry-go-round. Nonetheless, Dvořák finds it hard to tear himself away from the gay scene; in the C minor middle section he captures the very spirit of the fair, bustle, colour and mounting excitement. Finally he makes tracks for home when the 'cello theme returns, followed by other quiet variations, all in G major, which gradually dissolve in the romantic twilight of a summer evening. The symphony then ends in a rousing clatter on the full orchestra, the brass instruments being given many brilliant passages. Dvořák's 'Fourth' may not be his greatest symphony, but it has a compelling, individual fascination all its own.

On 15 December 1893 'From the New World' was given its first performance by the New York Philharmonic Orchestra under Anton Seidl. In Dvořák's own words, 'it created a furore'. But Europe took to it only a few years later. Today its worldwide popularity shows no signs of waning, even though the work itself has come in for considerable criticism as being mainly a succession of enchanting and virile tunes loosely strung together in patchwork style, presided over or helped out by a strongly rhythmic phrase bundled into each movement whenever Dvořák found himself wondering how best to proceed. Be that as it may, only a cynic can be deaf to the call of this warm-blooded music, so spontaneous it all sounds even in its moments of calculated joinery. It was composed in America during the three years from 1892 when Dvořák was Director of the New York National Conservatory of Music – a post held almost reluctantly by him at a salary of $15,000. The symphony was in fact a heartfelt greeting from the New World to his friends parted from him by circumstances and the ocean.

Much has been written about Dvořák's indebtedness to American influences in the actual music: to Negro Spirituals and the like. Seidl even went so far as to call it Red Indian music! But as Tovey points out: 'Dvořák's phrasing was primitive, Bohemian and childlike before he went to America.' These unsophisticated qualities shine through every bar of this music. And even if Dvořák himself, wanting success for it on both sides of the Atlantic, was given to such contradictory utterances as 'the influence of America must be felt', and 'forget all that nonsense about my using American national melodies', we still cannot fail to recognize the music as being primarily of Slavonic origin. While the G major flute melody in the *Allegro molto* first movement brings with it a breath of 'Swing low, sweet chariot' and while there are other suggestions of Negro melodies in the finale, Dvořák's fiery Czech imagination is always bursting ardently into flame. The 'New World' scherzo comes of the same stock as the *Furiant* of the D major symphony or the wilder Slavonic Dances; the *con fuoco* finale, its nine introductory bars, elementary enough to kindle the wrath of any pedagogue, set ablaze various themes and rhythms borrowed from the other movements, often in strenuous or playful combination. Even the *Largo* – composed, it is said, after Dvořák had been reading Longfellow's *Hiawatha* – here comes in for treatment far removed from its original nostalgic mood. Similarly the inspired modulatory harmonies (E major to D flat) with which this movement opens are subjected to yet another typically Slavonic onslaught.

In general terms much of the attractiveness of this symphony is due to its irregular features. While the music glows continually with the rich harmonies and orchestration of the nineteenth century, many of its themes, both primary and secondary, suggest some affinity with the pentatonic scale of five (variable) notes which together with its sixth note (the octave) has characterized folk-music in Asia, Africa, Europe and the Americas from time immemorial. Briefly, the 'flattened seventh' – D natural in the symphony's key of E minor – is a marked feature of many themes heard in all four movements. In Ex. 258 it is shown marked (x), not only in E minor, but also in other keys.

258

Ex. 258(*a*) occurs in the first bar of the symphony; (*b*) in the *Allegro molto* which follows; (*c*) in the *Largo* – whose tonality throughout is anchored to D flat major or to its minor mode, written as C sharp minor; (*d*) comes from the scherzo; (*e*) and (*f*) from the *Allegro con fuoco* finale. But for many cadences Dvořák is driven willynilly to use the conventional dominant-tonic progression of harmonies: B major with its D *sharp* to E minor or major, etc. And so in this recurrent conflict between naturals and sharps the listener's ear is continually attracted by the contrariness of it all. The final fifty bars of the symphony drop all references to the dominant-tonic cadences. The D sharp has disappeared. In this dramatic *tour de force* the symphony ends as loyal to its predominant D natural as it promised to be in its very first bar.

Most of the themes in the symphony are short in the limb, sometimes even bordering on the trivial, as for instance the unconscious 'three blind mice' phrases which in the finale are far too supine ever to run after the farmer's wife. All the same, Dvořák restores confidence with distinguished extended melodies both here and in the *Largo* movement. And in a happy phrase Šourek has compared the trio in the scherzo to a 'Bohemian inn where Schubert happens to be the guest'. The coda to this movement is a dramatic instance of Dvořák's calculated joinery, largely founded as it is on the symphony's 'presiding' rhythmic phrase, and on (of all things!) a *fortissimo* version for trumpets of the 'Swing low, sweet chariot' melody. Strangely enough, it all sounds right, wrong though some captious critics might argue it.

We come finally to the ubiquitous phrase which does duty throughout the symphony; a phrase set in a typically Slavonic rhythm (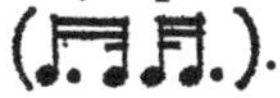). As shown below in Ex. 259, it acts in various capacities: (*a*) as a 'portent' in the *Adagio*; (*b*) as the main theme of the *Allegro molto*; (*c*) as a stormy irruption in the *Largo*; (*d*) a connecting link in the scherzo together with (*e*) further outbursts in this movement's coda; and then, loudly voicing its right to a certain supremacy, in diverse ways in the *Allegro con fuoco*. Here in its final appearance (*f*) it does battle with this movement's main theme and with two chords borrowed from the opening of the *Largo*.

259

Patchwork it may well be in places. But the warm humanity suffusing every bar and every orchestral touch in the 'New World' symphony will still continue to gladden the hearts of countless people the world over, come what may in musical fashions.

INDEX

The Companion to this volume

THE SYMPHONY VOLUME 2

Elgar to the Present Day

Essays on Elgar, Mahler, Nielsen, Sibelius, Roussel, Vaughan Williams, Rachmaninoff, Schmidt, Brian, Bax, Prokofiev, Rubbra, Walton, Shostakovich, Martinu and the Czech Tradition, Vagn Holmboe and the Later Scandinavians, and The Symphony in America.

By David Cox, Harold Truscott, Hugh Ottoway, John Manduell, Robert Simpson, Robert Layton, Peter Jona Korn.

THE PENGUIN GUIDE TO BARGAIN RECORDS

and

THE SECOND PENGUIN GUIDE TO BARGAIN RECORDS

Edward Greenfield, Ivan March, and Denis Stevens

'It would be hard to find better or more simply presented advice than Messrs Greenfield, March, and Stevens give us here' – *Musical Times*

'Genuine bargains are distinguished from what is merely cheap' – *Guardian*